THE QUIET EXTINCTION

The Quiet Extinction

Stories of North America's Rare and Threatened Plants

KARA ROGERS

**THE UNIVERSITY OF
ARIZONA PRESS**

TUCSON

The University of Arizona Press
www.uapress.arizona.edu

Printed in the United States of America
20   19   18   17   16   15   6   5   4   3   2   1

ISBN-13: 978-0-8165-3106-6 (paper)

Cover designed by Leigh McDonald
Cover photo: Mead's Milkweed (*Asclepias meadii*) by Jason Sturner

Library of Congress Cataloging-in-Publication Data
Rogers, Kara, author.
    The quiet extinction : stories of North America's rare and threatened plants / Kara
Rogers.
        pages cm
    Includes bibliographical references and index.
    ISBN 978-0-8165-3106-6 (pbk. : alk. paper)
    1. Endangered plants—North America.   2. Plant conservation—North America.
3. Endemic plants—North America.   I. Title.
    QK86.N66R64 2015
    581.68097—dc23
                                        2015005337

♾ This paper meets the requirements of ANSI/NISO Z39.48-1992 (Permanence of Paper).

Contents

Illustrations

THE QUIET EXTINCTION

Introduction

It is early January in Rocky Mountain National Park, and my husband and I are snowshoeing our way to Lake Odessa, more than 10,000 feet above sea level. The snowpack is already several feet up the trunks of the Engelmann spruce, subalpine firs, and limber pines that stretch above us. I feel a distinct sense of security as I stand among the trees, and the silence is profound. Neither of us speaks for fear that we will disturb the peacefulness here.

We are accompanied by a Clark's nutcracker, a lively, crow-sized, silver-and-black bird. It flits between the branches of the conifers, and each time it lands on a sprig of needles at the tip of a limb, a small puff of snow erupts into the air and showers gently to the ground. The sound it creates is like a muted rustling.

As we ascend in elevation, the trees thin out. Eventually they are reduced to stands of wind-tortured conifers that look as though they are trying to escape the brutal sweep of air, sprawling and tilting away from the direction of the prevailing winds. The trunks of some arch as if they have been pushed violently from behind.

The moment I cross the tree line I, too, adopt a peculiar lean, bending forward to counter the wind, which gives me an unwelcome shove. It howls at me, and when I turn to take in the panoramic view behind me, the wind bites at the back of my neck and nips through my insulated pants, numbing the backs of my knees. I look again at the trees, their bark twisted like a wet rag, cut with deep, arcing grooves. The feeling of desperation is inescapable.

The resiliency of conifers in the western mountains, the lushness of the temperate rain forest in the Pacific Northwest, the rolling green-leafed hills of New England—all give us the impression of healthy landscapes, places where plants are thriving. But green in the landscape obfuscates the struggle for life that currently defines the forests of North America, especially in the West, where over the course of the last several decades the rate of tree mortality in coniferous forests has doubled. The trees are dying faster than they are being replaced.

Few people are aware of the alarming rate at which plants are dying off or being extirpated from their natural environments in North America. The realization that the iconic trees of western North America are disappearing, that the great old-growth hemlock forests of the Appalachian Mountains are being felled by an invasive insect, or that the continent's native goldenrods, orchids, and other flowers are experiencing potentially disastrous declines has slipped silently by. The loss of plants and plant species has gone on quietly in North America. Many people may be surprised to learn that between 2000 and 2005 North America had the highest proportional loss of forest cover of all continents. Perhaps it would be surprising to discover too that several hundred species of plants in the United States and Canada are listed federally as endangered. Thousands more are thought to be threatened.

The need for greater awareness of not only the plants that are disappearing but also the places they are disappearing from, the factors that are contributing to their disappearance, and the consequences of their loss has grown dire. The value of what we stand to lose ecologically, culturally, and economically is significant. When we consider that the true economic potential of most at-risk species of native plants has yet to be realized and that native plants are embedded in our cultural and ecological heritage, the value of what we stand to lose becomes immeasurable.

Throughout much of the twentieth century, logging was considered to be the major threat to forests in North America. Today it is a significant issue only in certain regions. Other threats have emerged in its place, and many of them, unlike logging, are indifferent to the human-delineated boundaries that divide the land conveniently into public and private domains. The most insidious of these threats are climate change, disease, and pests, which have penetrated deep into our national parks and preserves, the last places where we would expect to find iconic species in decline. Each of these three threats is likewise indifferent to rural and urban boundaries, and they have combined in many cases with a localized event such as fire suppression or mining to decimate plant populations. Above all, however, the most common threat to plants is habitat loss. Deforestation, agricultural expansion, urban sprawl, and other forms of development have taken vital habitat away from plant species.

Unlike the introduction of an invasive fungus or insect or the manifestation of climate change, which can be difficult to detect, we have direct awareness and control over habitat loss. To act on that awareness and mitigate impacts on species requires some level of appreciation for plants and nature, though the challenge of cultivating or advancing such understanding is considerable. In Western countries in particular, many types of plants, trees especially, often are viewed as objects that stand in our way. Many people also are simply out of touch with the natural world, some to the point where they struggle to understand the land and its natural history.

But if we cannot conceive of the historical vastness of America's prairies, for instance, then we also are unlikely to be able to imagine the diversity or abundance of plants that once were a part of the prairies. Knowing so little about the natural environments that were inhabited historically by North America's indigenous peoples or more recently by European settlers represents a marked shift in baseline knowledge. The gaps that exist in our understanding of what nature once was and now is are sometimes immense, but it is crucial that we work to fill them, not least because in the case of plants we are easily deceived. Plants are extraordinarily diverse and resilient, but they also are sensitive to change. Their sensitivity often renders them vulnerable to decline, though in ways that are not necessarily obvious.

That plants still hold many secrets is evident in the wind-tortured conifers that survive on the high peaks of western North America's mountains. Some of those trees have been alive for as many as three or four centuries, enduring the wind and freezing temperatures during their entire existence. After a few moments in the fierce winds that scour the slopes where they grow, one wonders how anything can survive those conditions for a day, much less centuries. The extreme to which the trees have gone to find a niche in the world is amazing. But now, many of them are facing a threatened existence.

As we descend from Lake Odessa and reenter the shelter of the forest, we again cross paths with the Clark's nutcracker, as though it has been waiting for us. It is an intelligent and precocious bird. It knows, for instance, that we carry food, and because others of our kind have shared food with it, it thinks that maybe we will, too.

The bird is named for William Clark, who from 1804 to 1806 accompanied Meriwether Lewis on an expedition to the Pacific Northwest, exploring the newly purchased (1803) Louisiana Territory along the way. In 1805 the explorers crossed through Lolo Pass in the Bitterroot Range of the Rocky Mountains. It was a harrowing journey, one that traversed habitats that are difficult to imagine now. Fire suppression in the Bitterroot region has altered the natural processes that guide forest succession. Forests once dominated by ponderosa pine have been taken over by Rocky Mountain Douglas fir. The density of trees has grown, and the forests have become drier. They are primed for massive wildfire and for disease and insect infestation.

I think of those changes each time I see the silvery Clark's nutcracker, a species that along with so many other animals must find a way to survive as its habitat deteriorates. Many of North America's plants are facing similar situations, and some of those species, many perhaps, may decline significantly in the coming years. Those that are already endangered may go extinct.

Nature's recovery from the damage that we inflict upon it is not quick or simple, and maybe nature will not recover at all or not in ways that we might expect. And in the end, that inability to recover comes at our expense, for the great irony is that the land and its life, and how we use it and treat it, makes us who we are. In the loss of plants and inevitably the animals that rely on them, we stand to lose our own identity.

The value that we place on native plant species, from their aesthetic qualities to their agricultural and economic importance, forms a compelling and vital thread in the legacy of our natural heritage. Native plants, similar to traditions and relics, are pieces of that heritage. But unlike other objects and rituals, they are legacies of the land, and together with animals, they form inspiring and mysterious life mosaics. So an exploration of native plants really is much more than learning the science of species and their interactions. It is discovering what they mean to us, which in turn influences our personal and societal investment in their protection.

Interpreting that meaning is an unforgiving challenge, not least because nature means different things to different people. For some, it is an impediment to development or a thing to be exploited for monetary gain. For others, it is sacred and not to be disturbed. Between those two ends of the spectrum lies an array of value judgments and beliefs about nature. But a bit of common ground exists, even if small, and so there is hope for a healthy future, for nature and for our next generations.

The Quiet Extinction is focused on select plant species in the United States and Canada, north of the Gulf of Mexico. I have elected not to include species native to the Hawaiian Islands, as covering those species would fill an entire book in itself. The same can be said for the plants of

Mexico. Furthermore, due to the difficulty of condensing the sheer volume of information available, I have chosen to leave some of the finer clinical details of the featured continental species to the scientists' own words as they exist in the scientific literature. I encourage readers who may be looking for more information to sift through the scientific accounts cited in the endnotes section. It is my hope that this book, in relating the stories of plants on my home continent, will inspire greater environmental awareness and a deeper engagement with nature by people everywhere.

Whitebark Pine

When Scottish botanist John Jeffrey identified whitebark pine in British Columbia in 1851, he wrote of its location, "On the summit of a mountain near Fort Hope, Fraser's River." The peak rose 7,000 feet above the river valley, and it stood at the northern edge of the whitebark's prime habitat, a subalpine ecosystem in which the species was later discovered to play a fundamental role, shaping the character of the forest and supporting the survival of a diverse range of animals.

But in the century and a half since Jeffrey's observation, the subalpine whitebark ecosystem of British Columbia and the northwestern United States has moved steadily toward the brink of ecological disaster. Since at least the middle of the twentieth century, whitebark pine populations have been in decline. Their loss has been dramatic and complex, occurring as a result of a combination of factors, including infection with white pine blister rust. White pine blister rust is caused by the invasive Asian fungus *Cronartium ribicola*, which was introduced to eastern North America in the 1890s on infected white pine seedlings imported from Europe. It reached the West in the early twentieth century. Near the end of that century, a massive mountain pine beetle epidemic emerged in pine forests in western North America, which for whitebark compounded the effects of white pine blister rust. Those effects have been worsened further by fire suppression, which has restricted whitebark's regeneration in burned areas and led to its replacement by shade-tolerant species, and by climate change, a controversial and devastating factor at play in the loss of species.

In 2009 researchers found that within the Greater Yellowstone Ecosystem (GYE), which includes areas of southern Montana, eastern Idaho, and northwestern Wyoming, encompassing all of Yellowstone and Grand Teton National Parks—areas that constitute the southern portion of whitebark pine's range in the Rocky Mountains—some 46 percent of whitebark pine distribution was experiencing high mortality, while another 36 percent was affected by midlevel mortality.[1]

The suggestion was that within the GYE, whitebark would become functionally extinct, failing to perform its vital ecosystem services—providing food and shelter for animals and regulating the circulation of water through the subalpine—within five to seven years. That conclusion painted a dismal outlook for the species, and it was contentious, possibly overstating the urgency of the species' situation. Nonetheless, a year after the report was published, officials with the US Fish and Wildlife Service determined that whitebark pine was a candidate for protection under the Endangered Species Act.

Whitebark pine's quiet disappearance brings to mind Jeffrey's remarks, which forever linked him with the species in the history of botany in North America. It also brings to mind his demise. In 1854 Jeffrey disappeared silently, without a trace. His loss went unnoticed for some time. It was a sort of foreshadowing of whitebark's own erasure from the western landscape. The species' loss has been silent and difficult to comprehend, primarily because whitebark pine grows in the high, subalpine reaches of western North America's mountains, major portions of which lie within the boundaries of national parks and other protected areas. Relatively few people step foot in the high mountains, and so we might think that human activities have little impact there and that wild species must be thriving. Yet, whitebark pine, an iconic species, has been fading away slowly and steadily, with individual trees slipping inconspicuously to their deaths and the species on the whole creeping closer and closer to a threatened existence. Perhaps the tree's isolation on high mountain slopes prevented us from being able to know the true extent of its peril until recently. But we have also clung to a sense of security that comes with the knowledge that many of North America's wildest places are protected within preserves and national parks—a sureness that once seemed guaranteed but now is not.

Whitebark pine is a keystone species in the upper subalpine ecosystem, which means that its presence has inordinately large effects on the composite of life that has evolved with it and around it in the subalpine environment. It is also a foundation species, strongly influencing the stability of its community and ecosystem function. Those aspects of whitebark make it an especially interesting example of how the loss of a single species of plant has the potential to dramatically impact plant and animal communities and thereby alter the landscape. In regulating runoff from snowmelt, producing shelter and nutritious seeds, and preventing soil erosion, whitebark helps to retain moisture in subalpine soils, supports animal life, and keeps soil habitat intact for other plants and for animals. It creates the conditions necessary for the growth of other subalpine trees, including subalpine fir and Engelmann spruce. In the harshest conditions, at or near the tree line, whitebark initiates the formation of so-called tree islands, which often include other species of conifers.

We can begin to see, then, that without whitebark, the ecosystems that characterize the high elevations of mountains within its range in western North America would be very different from what they are today. The forest communities downslope and even plant communities in the mountain valleys, which benefit from the gradual, whitebark-regulated melt of snowpack in the subalpine, probably would be different, too. I hope that we never find out what those places would be like without whitebark. But the reality is that the species' decline is in process, and if greater awareness of its peril does not come, and if conservation efforts do not succeed, it is not inconceivable to think that very soon we will be left with only remnant populations in some parts of its range.

There is something in the shape and history of whitebark pine and in its ability to survive that makes it stand out from other species of trees. After seeing it for myself on Mount Rainier and in the Rocky Mountains, I came to realize that to truly appreciate its existence, to know its grace, and to understand the isolation that suppresses life on the high, subalpine slopes of North America's western mountains, I needed to think about the tree in that environment and the factors that give that environment its form. It is the circumstances—the geography, climate, and ecology—that have made whitebark pine what it is today. The same can be said of many other plants across North America's ecosystems.

It is easy to overlook the importance of a species like whitebark. There are more than one hundred different species of pines worldwide, about two-fifths of which are native to North America. Many of those trees historically were used by Native American peoples in the Northwest for both wood and nonwood products. Lodgepole pine, for example, was used in the framework of tipis, and the needles of white pines were used to make a tea to treat the common cold. Whitebark seeds once were a valuable source of food for indigenous tribes throughout the Northwest, including the Chilcotin, Flathead, and Kutenai peoples, who traditionally lived in the lands of southwestern British Columbia, Idaho, and Montana. Pines also generally

Figure 1. A subalpine ecosystem with whitebark pine habitat, Carthew-Alderson Trail, Waterton Lakes National Park, Alberta, Canada. (Photo credit: Jeremy D. Rogers)

have held a distinguished place in the history of botany, having been described through the work of Jeffrey and notable individuals such as Scottish botanist David Douglas and American botanists John Torrey and Asa Gray. But the ubiquity of pine trees and of conifers generally in the North American landscape, and the challenge of telling some of those trees apart, can make them seem as though they all are more or less the same and as though they are permanent, unchanging fixtures in the landscape.

The species of the pine genus (*Pinus*), however, are amazingly diverse, and the singularity of whitebark pine—seen in everything from its shape to the length and growth of its needles to its relationships with other species—reflects that. Its common name describes another unique trait, the pale color of its bark, a detail that German American botanist George Engelmann recognized when he gave whitebark pine its Latin name, *Pinus albicaulis*.

Whitebark pine is a major component of subalpine forests in the Coast Mountains in southern British Columbia, in the Pacific Northwest of the United States, and in the Cascade Mountains of both countries. It is also a major species in the Rocky Mountain subalpine ecosystem and in the high elevations of the Sierra Nevada of California. Its elevation range historically has extended from a couple thousand feet below to several hundred feet above the tree line. So in the Rocky Mountains, it may be found at elevations of 6,000 feet to more than 9,300 feet, while in the Sierra Nevada, it reaches points higher than 12,000 feet. In the Cascades of northern California, it is found between 7,000 and 9,500 feet, while from Oregon and north in those mountains, it grows from 5,400 to about 8,500 feet. It is also found in the Olympic Mountains of northwestern Washington, from 5,700 to 8,500 feet, and in the Blue Mountains of Oregon and Washington, between about 7,500 and 8,500 feet. At the far northern limit of its range, in central British Columbia and northern Alberta, it may drop to below 3,000 feet. For perspective, Mount Rainier, the highest point in the Cascade Mountains, stands at 14,410 feet, and Mount Whitney, at 14,494 feet in the Sierra Nevada, is the highest peak in all the Pacific Coast ranges. On both mountains, whitebark pine is one of the last trees that hikers pass as they ascend clear of the tree line and into the granite barrens.

In the lower parts of its elevation range, whitebark pine often grows as a component of mixed stands alongside species such as lodgepole pine (*Pinus contorta*), Engelmann spruce (*Picea engelmannii*), coast Douglas fir (*Pseudotsuga menziesii* var. *menziesii*), Rocky Mountain Douglas fir (*Pseudotsuga menziesii* var. *glauca*), subalpine fir (*Abies lasiocarpa*), mountain hemlock (*Tsuga mertensiana*), and foxtail pine (*Pinus balfouriana*). Which trees it shares habitat with depends on climatic and geographic factors, as well as other characteristics unique to the various mountain ranges. As elevation increases, whitebark generally emerges as the dominant and in some instances the only species of tree. In forests dominated by whitebark pine, the understory is often discontinuous and includes plant communities made up variously of grouse whortleberry (*Vaccinium scoparium*), common juniper (*Juniperus communis*), Oregon boxwood (*Paxistima myrsinites*), pink mountain-heath (*Phyllodoce empetriformis*), subalpine fleabane (*Erigeron peregrinus*), and various types of sedges and grasses. The particular species found with whitebark in a given area are again a function mainly of geography and climate.

Whitebark pine's resiliency and talent for survival are most evident at or near the tree line, where in winter the frozen landscape seems to still time and magnify the sense of solitude. It is in that environment where the tree carves out a most remarkable existence, growing in isolated stands of creeping, stunted conifers known as krummholz. The first whitebarks that fixed themselves in my memory were those in krummholz formations, and I remember being struck by a suggestion of infinite patience in their eternal tolerance of the short growing season, despite their wanting and constrained appearance. In the high elevations, where krummholz trees are found, everything seems smaller than it might otherwise be, dwarfed by the wind, by the limited resources, and by the relative immenseness of the mountains.

The elements and time traditionally have dictated life in the subalpine and alpine environments. The rate of growth and time to maturity for whitebark pine at tree line is appropriately glacial. A four-hundred-year-old whitebark at such elevations may measure a gaunt half foot in diameter, and its shape is often sprawling, with limited vertical growth and relatively extensive horizontal growth in the direction away from the wind. Whitebarks at tree line look weather-beaten, with twisted trunks, sometimes thinly split or grooved bark, and limbs bent and curled into improbable formations during their growth over decades of exposure to the wind. The difference is stark when those trees are compared with their counterparts growing farther downslope in the upper subalpine below the tree line. Whitebarks that grow in those areas and in areas protected from the wind may reach 50 or 60 feet in height and 1 or 2 feet in diameter. They also often have fairly straight trunks and branches that stretch upward. But even in the lower parts of its elevation range, whitebark, like many other conifers, still requires decades to reach full size.

Whitebark pine's slow rate of growth means that the tree is slow to mature reproductively. Female whitebarks require at least two decades before they begin to produce cones, and initially they do not produce very many cones. Full cone production is not achieved until the tree has become moderately large, as cone crop size appears to be related to canopy dimension. The timing of cone production is not very different from that of other subalpine conifers, however. Engelmann spruce, for example, begins cone production after one to four decades of growth, with full seed production coming between one and three centuries of age. Relative to Engelmann spruce and species such as Rocky Mountain Douglas fir and subalpine fir, whitebark is a shorter tree, with a canopy shape that may actually be broader.

Whitebark pine is one of five species of stone pines that exist in the world, and it is the only species of the group that occurs in North America. Stone pines, which are named for the large, heavy seeds they produce, are further distinguished by having five needles per needle cluster (fascicle) on their branches and by the way in which their cones remain closed after reaching maturity, a phenomenon known as indehiscence. Examples of other species of pine that produce cones that require an external factor to coax them open include lodgepole pine and jack pine (*Pinus banksiana*), both of which produce serotinous cones, which are sealed by a resin and open upon exposure to temperatures high enough to undo the seal. Other species of pine in North America produce dehiscent cones, or cones that open more or less on their own. Pines with dehiscent cones release winged and usually small, light seeds, which are readily dispersed by the wind (the enormous winged seeds of sugar pine, *P. lambertiana*, are a notable exception). Whitebark, by contrast, has wingless seeds, so rather than relying on the wind to carry its seeds, it depends on an animal for seed dispersal. That animal is a bird, and apparently Jeffrey witnessed it himself in the vicinity of whitebark pines, "depriving them of nearly all their seeds."[2] He described the bird he saw as *Corvus columbianus*, a species also known as Clark's nutcracker (*Nucifraga columbiana*).

Whitebark pine relies almost exclusively on Clark's nutcrackers for its regeneration, which means that most whitebark trees in North America have been planted by one of these birds. Clark's nutcracker, however, does not depend solely on whitebark seeds as a source of food. Rather, it is specially adapted for scavenging, transporting, and caching the seeds of various types of pines, including limber, pinyon, and whitebark.

One of Clark's nutcracker's more marked anatomical traits is its long, pointed beak, which is such a prominent feature that William Clark, who saw the bird in 1805 during his expedition with Meriwether Lewis to the Pacific, mistakenly took the animal to be a kind of woodpecker. The bird uses its beak to pry open the cones of conifers and to test pine seeds for

Figure 2. Whitebark pine (*Pinus albicaulis*) cone. (Credit: Kara Rogers)

ripeness and edibility. Another key, though less obvious, adaptation of Clark's nutcracker is its sublingual pouch, which lies beneath the tongue and allows the bird to transport seeds in bulk to distant caching sites. A Clark's nutcracker can pack many whitebark seeds into its pouch at one time, and it may carry the seeds as far as 15 to 20 miles from collection sites. Its seed-carrying capacity enables a single bird to cache tens of thousands of seeds of various species of pines in one season. It buries the seeds in numerous caches, which the bird often creates at the bases of trees or rocks and in open areas (such as burned lands); these caches are distributed generally in the lower elevations of each bird's home range. The bird relies on its remarkable memory to locate caches to retrieve seeds throughout winter. But every now and then it forgets or simply does not return. Out of those forgotten and lost stores, in which conditions typically are ideal for seed germination, sprout new generations of seedlings.[3] The process of regeneration is a gradual one. Whitebark seeds buried in a forgotten cache lie dormant for two or more winters before germinating.

Seed caching by Clark's nutcrackers may impact the genetic history of whitebark populations. Perhaps most notable is a possible homogenizing effect on the genetic diversity of whitebark arising from the way in which the birds distribute seeds across wide geographic areas.[4] Such seeds typically are full siblings and thus are closely related genetically. In conifers, homogenization may be further associated with pollen movement, in which pollen is swept by the wind over significant distances, resulting in the transfer of genes from one population to another. At the species and population levels, whitebark exhibits relatively low levels of genetic variation. Among individual whitebark trees, however, there exists marked genetic diversity. Some trees possess genetic resistance to the destructive white pine blister rust, which has important implications for the species' restoration.

The influence of Clark's nutcracker may extend to the overall evolution of whitebark pine, with effects particularly on the tree's shape and certain characteristics of its cones. Most whitebarks, similar to other white pines, have branches that project upward, with cones positioned horizontally, which makes them obvious to birds flying overhead. Whitebarks also produce cones that, when mature, despite being indehiscent, have relatively fragile scales that can be broken or torn off by Clark's nutcracker to expose the seeds inside.

It may be the way in which the cones are disassembled on the branches by the bird that explains the frequent absence of cones on or beneath a whitebark, a phenomenon that Jeffrey noted. Often, fragments of cones and cone scales can be seen scattered about near whitebark pines. The generally inaccurate notion that the tree might normally produce very few cones is compounded in years of actual low cone production and by the behavior of squirrels, namely, North American red squirrels (*Tamiasciurus hudsonicus*) and Douglas squirrels (*T. douglasii*), which collect whitebark cones and store them underground in one or several middens, further keeping the tree's cones out of view. Interestingly, the seeds that the squirrels cache do not play a role in the tree's regeneration, because the seeds either remain trapped in the cones or are dug up and eaten. Stray seeds on the ground or lingering in partially opened cones may be eaten by species of birds that otherwise are unable to open the cones themselves. Examples include mountain chickadees (*Poecile gambeli*, or *Parus gambeli*), pine grosbeaks (*Pinicola enucleator*), and woodpeckers (*Picoides* spp.).

Clark's nutcrackers are well rewarded by whitebark for their seed-caching efforts. While it takes just over two years for the seeds of whitebark pine to mature (a two-year reproductive cycle, typical of pines), once fully developed they are larger than the seeds of most other conifers that grow in subalpine ecosystems. Each whitebark seed measures about one-third of an inch in length and one- to two-tenths of a gram in weight. More than half of that weight consists of fat, making the seeds an especially rich source of calories.

Squirrels, black bears (*Ursus americanus*), and grizzly bears (*U. arctos horribilis*) also benefit nutritionally from whitebark seeds. During winter, squirrels retrieve seeds from their middens. Several species of birds, including mountain bluebirds and northern flickers, nest in whitebark trees. While whitebark pines are not an important source of food for bighorn sheep or white-tailed jackrabbits, those animals may benefit from the shelter provided by whitebark thickets and islands.

The ability of whitebark to support those species, as well as carry out its other functions in the subalpine, is deteriorating. Whitebark pines that become infected with white pine blister rust will eventually die. While whitebarks in the southern Sierra and Great Basin have very low rates of infection or no infection, the rate of infection elsewhere is higher. For example, in the greater Yellowstone area, the infection rate has been about 20 percent on average, while higher rates have been found in the Northern Continental Divide Ecosystem, which includes the Bob Marshall Wilderness, Glacier National Park, Flathead National Forest, and Blackfeet Indian Reservation in Montana and the adjacent Waterton Lakes National Park in Canada. In some whitebark forests within that range, infection rates have reached nearly 100 percent. In general, the disease is most severe in the more northern parts of whitebark's range, where the climate tends to be more humid and cool in late summer, conditions that are ideal for the production and transport of fungal spores to new host trees, where the spores can then germinate.

The *Cronartium ribicola* fungus has an exceedingly complex life cycle, with two hosts and five different spore types, two of which are involved in wind-disseminated transfers between hosts. In the spring, aeciospores are dispersed from pines and carried on the wind (sometimes over long distances, up to 310 miles) to alternate hosts, which most often are deciduous gooseberry or currant (*Ribes* spp.). The rust forms tiny, orange-colored colonies on host plants. In late summer, it releases basidiospores, which are dispersed by the wind and infect nearby white pines. Because alternate host plants shed their infected leaves, they are not adversely affected by the fungus. On whitebark and other white pines, however, the fungus works its way through tiny openings (stomata) on the surface of pine needles. It then grows deeper into the tree, propagating through needles and into twigs, then into branches, and ultimately into the trunk. It penetrates as far as the cambium, a thin layer that rings the tree, to the inside of which lies the water-conducting xylem and to the outside of which lies the nutrient-conducting phloem. In the case of mature trees that become infected, the killing process can take many years, with branches dying off one by one. As branches die, part of the canopy of the tree dies, too. Since female cones develop in the canopy, increased canopy death causes infected trees to become reproductively dead. In the last stages of disease, the rust grows into the stem and the tree is girdled, cutting off the flow of nutrients and causing the top portion of the tree to die. Young trees, because of their small size, succumb to the disease more quickly than mature trees.

While the fungus is quick to cause the deaths of young whitebark pines and also attacks mature trees, the mountain pine beetle (*Dendroctonus ponderosae*) seems to favor older, more mature whitebarks, which have larger diameters and thicker, more insulating bark. As a type of bark beetle, the mountain pine beetle reproduces only beneath tree bark. It specifically targets several different types of pines, including lodgepole pine, jack pine, and whitebark. It chews its way through the bark and into the inner vascular tissues of the tree, stopping only once it reaches the phloem. In the process, the insect, which is about the size of a grain of rice, leaves piles of sawdust to form on the ground at each tree's base—a telltale sign that a bark beetle is at work. In severe mountain pine beetle epidemics, such as the infestations of the 1930s and 1940s, the 1970s, and the late 1990s through 2010/2012, a tree may become riddled with hundreds of bore holes.

Mountain pine beetles are active in the summer months. At that time, adult beetles emerge from dead or dying host trees and leave those trees in search of new, living hosts. When a female mountain pine beetle has located a suitable host tree, she releases a pheromone that attracts male beetles and more female beetles. Meanwhile, as the beetles chew their way into the vascular tissue, the tree attempts to plug the bore holes by exuding sap, creating what are known as pitch tubes. As infestation progresses, females mate with males and then lay their broods of eggs within galleries carved into the host's inner wood. That activity and the introduction of blue-stain fungus (*Grosmannia clavigera* and *Ophiostoma montium*), which is carried by adult beetles and is used as food by larvae, ultimately disrupts the transport of nutrients through the phloem. The tree begins to die from the inside out. By winter, the developing larvae are tucked safely in their galleries beneath the bark, feeding on the tree and on the blue-stain fungus, which by that time has become thoroughly embedded in the phloem, completely impairing its flow. So long as winter temperatures do not become too cold, the larvae will survive and emerge the following summer to begin the cycle of infestation anew. Once they have reproduced, adult beetles from the previous generation disperse and die. As the tree that has been the victim of infestation begins to die, too, the remaining chlorophyll in its needles breaks down, causing the needles to turn from green to red, a color that has become disconcertingly common in western North America's pine forests.

The mountain pine beetle outbreak that began in the 1990s in the West and that only recently began to slow was large and severe. In British Columbia, it was estimated to have been the largest outbreak on record.[5] There, the beetle infested an estimated 40 million acres of pine forest. The annual acreage of trees under attack by the insect climbed into the millions between 2001 and 2002 in the United States and eventually peaked just short of nine million in 2009. Climate warming was cited repeatedly as a significant contributing factor in the severity of the outbreak. In particular, unusually warm winters and limited or low snowpack in the mountains in winter and early spring created conditions that researchers concluded were amenable to the production by some broods of two generations, rather than one, each summer.[6] Some beetles began laying one batch of eggs in late spring or early summer and a second in late summer, which had been the traditional time of egg laying. The advance in egg production could have led to a sizeable increase in the beetle's numbers.

Although there has been some speculation that climate warming allowed mountain pine beetles to penetrate into unprecedentedly high elevations, many researchers now think that the beetles were present historically at those elevations but that they had existed only in small numbers and had affected only certain species of white pine. Whitebark pine probably was not one of those species, which means that even though it may have shared its elevation range with the beetle, it did not evolve defenses against beetle attacks. In the Rocky Mountains, the susceptibility of whitebark to mountain pine beetle infestation probably has been aggravated by recent changes in climate. Climate data gathered by the National Oceanographic and Atmospheric Administration (NOAA) have shown that the amount of snowpack in most mountain locations in western North America has been decreasing and that snowpack has been melting slightly earlier in the year, triggered by unusually warm spring temperatures.[7] Reduced snowpack and early melting of snow, which have been linked to cyclical climate variations (such as the Pacific Decadal Oscillation and the El Niño–Southern Oscillation), have been further associated with increased drought in western regions. Whitebarks that have been weakened by drought may succumb more quickly to white pine blister rust and beetle infestation than trees unaffected by drought.

Climatic factors also appear to be linked to fire in western forests. Drought, for example, primes trees for fires that are capable of devastating large areas of forest. The combination of drought and high tree density from fire suppression sets the stage for especially large and intense wildland fires. Fire suppression implemented widely in North America in the early twentieth century allowed some forests to become especially dense and thereby raised the risk of massive stand-replacing fires. Those types of fires, over which humans have little control, began to affect some parts of the American West in the late twentieth and early twenty-first centuries.

The regeneration of whitebark in some communities, particularly in the Rockies, depends on some level of natural fire activity. In centuries past, fire in whitebark ecosystems appears to have occurred at varying intervals and with varying intensity, depending on the amount of fuel and moisture in the ecosystem and on other factors, such as wind. Based in large part on fire scars recorded in the growth rings of trees, fires that affect whitebark ecosystems can be divided into three basic types: low-intensity burns, which appear to be more typical in the southern portions of the tree's range in the Rocky Mountains, and mixed-severity and high-severity (stand-replacing) burns, which appear to be more common in the tree's northern range. Low-intensity burns are recurring, nonlethal burns that clear out fire-sensitive species, such as spruce and fir. Whitebark's relatively sparse crown and thick bark are thought to enable it to withstand low-intensity fires. Mixed-severity fires, which range in intensity from nonlethal understory fires to tree-killing fires, occur about every 50 to 300 years, cover areas of about

2 to more than 70 acres of land, and create gaps in the forest canopy, providing openings for Clark's nutcrackers to plant seeds in caches. The most intense burns occur over long time scales (generally about every 250 years), originate in low-elevation forests and tend to move upslope, and kill upward of 95 percent of trees.

Large swathes of burned land, which lack shade and competitor species, provide an ideal situation for whitebark regeneration. They also provide potential seed-caching territory for Clark's nutcrackers, and because the birds carry whitebark seeds farther than the wind can carry the seeds of potential competitor tree species, whitebark generally has an advantage over other conifers when it comes to recolonizing burned subalpine environments. In fact, unlike some other types of trees found at subalpine elevations, whitebark is a pioneer species—in the subalpine environments across its range, it is among the first conifer species to populate burned areas. It moves in and creates microsite conditions, such as small patches of shade and possibly moisture, that enable the survival of other conifers and plants, thereby facilitating community development. Historical, natural fire regimes, by varying in intensity, also gave shape to a mosaic of different-aged forests on the landscape and thereby helped to maintain biodiversity.

As long as healthy seed sources are available, high-intensity fires should have little adverse effect on whitebark pine. Even though adult trees would be killed by fire, they would be replaced by trees that emerge from caches. But the decline of seed sources as a result of mountain pine beetle infestation and particularly infection with white pine blister rust has made tree-killing fires an issue in whitebark communities. In areas where whitebark seed sources are mostly dead due to the reproductive death of trees as a result of white pine blister rust, tree-killing or stand-replacing fires could devastate the species' populations.

The extensive ecological roles fulfilled by whitebark pine mean that its loss would have significant ramifications in the mountain ecosystems across its range.[8] Alterations in the shading of snow and soil retention in the montane environment would affect snowmelt and stream flow. Climate change could cause other species to move upslope and outcompete whitebark, resulting in the replacement of whitebark communities by other forest types. Populations of Clark's nutcrackers might be affected, as they could suffer small declines and alter their habits in affected areas. Since Clark's nutcrackers help to disperse the seeds of other species of white pine, decreases in their populations or changes in their behavior could affect populations of those pines as well.

The decline of whitebark could also have significant impacts on humans. People who depend on downstream water supplies could see streams dry up earlier in the summer or fall, since spring snowmelt would be hastened as a result of combined whitebark losses and climate warming. Further combined with preexisting drought conditions, such hastening could be particularly damaging to agriculture and could limit seasonal water resources. Whitebark communities also help to maintain biodiversity, ensuring their own health and the vitality of the landscape. In the disappearance of whitebark, Native Americans would lose a traditional source of food and medicine that once held a role in their cultures.

The disappearance of whitebark also would affect the aesthetic appeal of the upper limits of the subalpine and tree line. Twisted, gnarled trees and low-growing krummholz formations at or near the tree line provide a unique visual aspect to the landscape. Many people who visit national parks and other nature destinations in the northwestern United States and in southwestern Canada are rewarded with spectacular views, of which subalpine forests are major components. The pervasive red coloring of dying trees and the gray skeletons of dead ones have already affected the scenic nature of pine forests in those regions. The impact on aesthetics also extends to the spiritual qualities of the subalpine, where ancient trees have stood for hundreds of years like guardians over the high mountains and have provided inspiration for humankind.

The survival of whitebark pine across many parts of its range now depends on restoration, plans and activities for which have been developed and implemented by researchers, land managers, and resource specialists with the support of federal agencies and nongovernmental organizations. The successful restoration of whitebark requires an approach that centers on four principles: promoting resistance to white pine blister rust, conserving genetic diversity, saving seed sources, and employing restoration treatments. Actions that support those principles include assessments of species status and trends, population monitoring, activity planning, reduction of disturbance impact, seed collection, growth and planting of rust-resistant seedlings, conservation of seed sources, the creation of habitat conditions to encourage whitebark regeneration, sampling to evaluate the success of restoration treatments, and research to refine the techniques of restoration.

The restoration approach for whitebark pine, which was developed by research ecologist Robert E. Keane and colleagues, is designed to address issues over a range of spatial scales, from tree and stand to landscape, forest, region, and full range.[9] Within that framework, areas will be prioritized for restoration, with highest priority given to places heavily affected by blister rust. Other prioritized areas include those with high mortality from mountain pine beetle infestation and those affected by advanced succession, as well as those only moderately impacted by blister rust. The approach is also grounded on key concepts in restoration that take into account whitebark pine's historical range and variability, its physiological constraints, and its community, including interactions between species and between species, physical disturbances, and climate change. The latter allows for planning into the future, which for whitebark may mean assisted migration, in which seedlings from warmer, lower latitudes are planted into the upper subalpine environment at more northern latitudes within the species' range to facilitate survival in a warming climate.

Whitebark pine's genetic diversity is especially important to its restoration. To create a record of genetic diversity, seeds are collected from trees in areas affected by blister rust and from trees in unexposed areas across the species' range. Archived seeds are used for the creation of viable seed inventories, which provide for worst-case scenarios, such as the elimination of whitebark from parts of its natural range. Seed collections also serve as the foundation for the development of planting stock and potentially could be used for the direct seeding of whitebark onto treated or burned sites.

To identify potential planting stock, seeds undergo rust screening. The major centers for rust screening are the Coeur d'Alene Nursery in Idaho and the seedling nursery at the Dorena Genetic Resource Center in Oregon, both managed by the USDA Forest Service. Seeds from potentially rust-resistant whitebark trees are grown into seedlings. The seedlings are exposed to high levels of spores of the *Cronartium ribicola* fungus. The parent trees that produced seedlings that are asymptomatic following exposure and that show resistance reactions are then used as seed sources. Seedlings grown from rust-resistant parents are planted in areas that have been prepared with appropriate treatments, examples of which include the removal of competitor species and the creation of seed-caching habitat for Clark's nutcrackers.

Once in the ground, seedlings must be monitored closely to determine whether the applied restoration treatments have been successful. Monitoring may include assessments of seedling survival over time and of blister rust infection in heavily affected areas. In addition to the monitoring of new populations, rust-resistant parent trees that serve as seed sources need protection against mountain pine beetle attack.

The story of whitebark pine as a declining species is one among many in North America. American chestnut (*Castanea dentata*) once dominated the forests of eastern North America. It was that region's giant sequoia. Its nuts were a source of food for animals, and its wood, which was highly resistant to decay, was harvested by humans for a variety of purposes. But sometime around the start of the twentieth century, the fungus *Cryphonectria parasitica* arrived in the United States. It eventually decimated the American chestnut, and only through intense and arduous restoration efforts has that iconic tree begun to rebound from the precipice of extinction.

Across North America, forests are in trouble. Between 2000 and 2005, North America lost nearly 114,000 square miles of forest cover—an area equivalent to the size of the state of Arizona. It was the greatest expanse of forest cover lost on any continent during that time, and it accounted for an astonishing 5 percent of total forest cover worldwide and almost 18 percent of North America's total forested land.[10] Much of that area had been lost to logging in the southeastern United States and southern Canada and to wildfire in northern boreal forests. At the same time, outbreaks of insects, such as budworms, Asian longhorned beetles, emerald ash borers, and spruce pine beetles, and of diseases, such as Dutch elm disease and sudden oak death, were wiping out trees in forests across the United States and Canada. And on top of it all, since the 1980s and 1990s, western North America's old-growth forests have experienced a doubling in rate of tree mortality, an effect associated with drought and regional climate warming.[11]

Water deficits and climate change, along with logging and wildland fires, are major factors underlying the loss of forests in North America. While logging was once at the forefront of environmental concern and remains a contentious issue, it has been overshadowed by climate change. A rise in the frequency and intensity of wildfire and the noticeable effects of drought have been linked to climate warming. These factors have coalesced to cause the deaths of millions of trees, and in the process they have set in motion ecological changes so vast and involved that scientists struggle to understand them, much less anticipate their outcomes. Nevertheless, we can be sure that the consequences of losing our forests in North America will be far-reaching. And it is not just trees that are disappearing. About a thousand species of plants in the United States and Canada are recognized formally as being threatened or endangered, and several thousand more are at risk of soon reaching threatened status.

The loss of plants arguably is one of North America's greatest ecological travesties. The variety of species that have been affected and the massive areas of land that have been altered are shocking. But the real tragedy is that few people are aware that the region's iconic trees and beautiful plants are disappearing and that they are doing so rapidly. When people think of the extinction of plants, they likely think of rain forests in the tropics, not of the forests and flowers in their own backyards or in their country's protected parks. Yet in North America, just as in the tropics, plants are fundamental components of the land and the region's history. In their loss, we stand to lose not only nature but also our continent's natural heritage—the traditions, culture, and legacy of its people.

Fraser Fir

When it comes to traditions, few rival the Christmas tree in universality. Regardless of whether or not one chooses to celebrate the holidays with it, the Christmas tree is a part of many people's lives come the wintertime holiday season. The tradition began sometime around the sixteenth century, when various customs, such as the honoring of sacred trees in Scandinavia and the placement of an undecorated Yule tree at the entrance to or within the house in Germany, converged. The distinguishing features of the Christmas tree were its decorations and its use specifically for Christmastime celebrations, which marked a divergence from past customs.

Long before there was the Christmas tree there were simple evergreen boughs. Used to observe the winter solstice by ancient Romans, ancient Egyptians, and ancient Celts, evergreen boughs were, to many, reminders of eternal life. To some, the continuance of life was captured symbolically in an evergreen wreath, to which there was neither beginning nor end. For most, the presence of boughs of evergreens in the home was cause for celebration. Some ancient cultures appear to have gone beyond branches, too, making entire evergreen trees a part of their winter solstice celebrations. Out of those traditions, presumably, emerged the ideas that inspired the eventual development of the Christmas tree so many centuries ago.

Some of the most ancient traditions, particularly the use of holly boughs, are still practiced today. They have not yielded to time, in part because evergreens, including the Christmas tree, are not tied to any one meaning. In other words, although given specific interpretations in Christianity or other religions, evergreens basically are devoid of religious bearing. They simply are pieces of nature, which makes them accessible to anyone.

The Christmas tree has become so integral to tradition for many people that it is extremely difficult to imagine what the holidays would be like without it. In North America, one may be greeted by that same feeling of incomprehensibility when trying to imagine, for example, the Rocky Mountains or the Appalachian Mountains without their cone-bearing evergreens, among which are the wild relatives of cultivated Christmas trees. Both the absence of the Christmas tree and the absence of wild conifers blanketing the mountains seem unlikely scenarios, impossible even, until one begins to think about the future of North America's coniferous ecosystems in the context of climate change and invasive pests.

The Christmas tree and the conifers of North America are connected: they essentially are one and the same. They also are components of our natural heritage. We enjoy them today because the traditions and legacies associated with them continue to be handed down to new generations. In the case of the Christmas tree, people have adopted the annual practice and developed an appreciation for it from their families and culture. In the case of wild conifers, appreciation has been enabled by the perpetuation of their existence, which has become increasingly dependent on their conservation as a result of various threats impinging on coniferous ecosystems. But their conservation does not occur in isolation, away from their cultivated counterparts. For while the Christmas tree tradition may seem entirely separate from any consideration of nature's welfare, it has very real consequences for conifers in the wild.

People have made a ritual of selecting Christmas trees, bringing them into their homes, and decorating them. The ritual becomes a part of many people's lives beginning in childhood. From a very early age, many people ascribe a kind of meaning to the trees that is unique among plants, and they form a connection with conifers that endures throughout their lives. While it may not be acknowledged and while it may not be common to every person, the connection can be remarkably strong owing to an emotional bond. A perfectly A-shaped spruce or fir in a mountainscape might bring back happy memories of the holidays, even for those who choose not to celebrate the season with a Christmas tree. The simple act of recalling those memories and the positive emotions that likely are associated with them might add to the experience of seeing conifers in the wild. The connection formed emotionally from that experience has lasting significance when it comes to the conservation of a species like Fraser fir (*Abies fraseri*), which is one of the most popular species of conifer cultivated for use as a Christmas tree in North America.

Fraser fir is at risk of extinction in its native habitat in the Southern Appalachian Mountains of North Carolina, Tennessee, and Virginia. There is great disparity in its abundance, as its cultivated form far outnumbers its wild form. Indeed, the success of the former easily belies Fraser fir's vulnerability in the wild. But cultivation has also enabled it to become thoroughly embedded in the Christmas tradition in North America. As a key part of that tradition, it has fed into Americans' familiarity with conifers, and in a physical and an emotional sense, it has connected people with nature. In doing so, it has enhanced the significance of our customs and natural heritage, and it has revealed that even in the spectacle of Christmas and all the materialism that comes with it, the bond between human and nature endures. That connection implies a responsibility to see to the perpetuation of Fraser fir's well-being, particularly in its native habitat.

Modern knowledge of Fraser fir began with its discovery in the late 1780s in the Southern Appalachians of North Carolina by Scottish botanist John Fraser, for whom it is named. Fraser fir has since been found to exist in only six populations: on Mount Rogers in southern Virginia; on Grandfather Mountain, Balsam Mountain, and the Black Mountains in North Carolina; and on Roan Mountain and the Great Smoky Mountains marking the border between North Carolina and Tennessee.[1] Each of these areas is characterized by a cool climate with frequent fog and heavy snowfall, which contribute large amounts of moisture to the habitat. Moisture and cool temperatures are key to Fraser fir's ability to thrive. Above 4,500 feet, which marks the lowest extent of the species' elevation range, the mountains can receive anywhere from 75 to 100 inches of precipitation in a year. At the highest elevations, around 5,900 feet and above, summer temperatures generally average a cool 60°F. In regions above 6,200 feet, Fraser fir is the dominant conifer. In the highest portions of the Southern Appalachians, such as on Mount Mitchell in the Black Mountains (which at 6,684 feet is the highest peak in eastern North America), it occurs in pure stands. In the lower parts of its range, between about 5,400 and 6,200 feet, it grows in stands with red spruce (*Picea rubens*), giving rise to the red spruce–Fraser fir ecosystem.

The elevations where the spruce-fir ecosystem exists overlap to some extent with the higher boundaries of deciduous species such as beech, birch, maple, mountain ash, pin cherry, and serviceberry. Yellow birch (*Betula alleghaniensis*) is sometimes so abundant as to be considered a third major overstory component. Thus, the lower elevations of spruce-fir forests, in also serving as homes for stands of deciduous species, are complex ecosystems with a great diversity of plant and animal life. In the undergrowth of these forests one is likely to find a number of different types of mosses and ferns;

Figure 3. Red spruce–Fraser fir forest, Roan Mountain. (Photo credit: Kevin M. Potter, North Carolina State University)

a variety of shrubs, including mountain cranberry and blackberry; and wild-flowers such as lily and mayflower. As many as one dozen different species of plants may be endemic to the spruce-fir ecosystem, as are multiple species of insects and other animals, including the Carolina northern flying squirrel (*Glaucomys sabrinus coloratus*), the magnolia warbler (*Dendroica magnolia*), and the spruce-fir moss spider (*Microhexura montivaga*).

Fraser fir, then, is a sort of magnet, whether for biodiversity or for orna-ments, and it seems made to fulfill those roles. It is, interestingly, unsuited for the uses for which one might think a tree would be ideal in human culture. The relatively poor material strength of its wood renders it of little value to the timber industry, and the difficulty of cultivating it at elevations below those of its natural range and its susceptibility to pest infestation limit its use as an ornamental.

In suitable elevations in North Carolina, Fraser fir has been cultivated since the mid-twentieth century with great success for the Christmas tree industry. In recent decades, its popularity among consumers in the United States has grown, primarily because more people have become aware of its fine Christmas tree qualities, which include its conical shape, pleasing fra-grance, good needle retention, and soft needles. Promotion of the sale of

Fraser fir in North Carolina has led to significant expansion in the tree's cultivation there: today some fifty million Fraser firs populate roughly 25,000 acres on tree farms in the state. Annual revenue from Christmas tree sales in North Carolina has hovered around $100 million, with Fraser fir accounting for more than 95 percent of that figure. Only in Oregon does Christmas tree production exceed those numbers. There, in 2007 about 6.9 million Christmas trees were harvested off 61,850 acres of land devoted to tree farms. The sale of those trees amounted to $109 million, 90 percent of which came from the sale of Douglas fir (*Pseudotsuga menziesii*) and noble fir (*Abies procera*).[2]

Douglas fir and noble fir are found throughout much of the Pacific Northwest and parts of California. While the fir varieties known as Rocky Mountain Douglas fir (*Pseudotsuga menziesii* var. *glauca*) and coast Douglas fir (*P. menziesii* var. *menziesii*) are threatened in limited portions of their range (specifically in Nevada), both Douglas and noble fir are otherwise considered to be at low risk of extinction. Fraser fir, on the other hand, is listed as a species of concern federally, as threatened in Tennessee, and as vulnerable by the IUCN Red List of Threatened Species.[3] As one of the two key species that define the spruce-fir ecosystem of the Southern Appalachians, Fraser fir's vulnerable status means that the whole of the ecosystem may be in jeopardy, too—a risk further increased by the fact that red spruce is also on the decline. The ecosystem is considered to be one of the most endangered in the United States.

The early European settlers who first arrived in the Southern Appalachians probably took one look at the steep terrain where red spruce and Fraser fir grow and turned around. To log the high slopes would have been impracticable then. But by the early twentieth century, with the development of mechanized logging and the penetration of railroads into the southern mountains, the coniferous forests of the region suddenly were within reach of the timber industry. Logging and slash fires, some of which may have escaped into the high elevations from deciduous forests downslope, impacted populations of Fraser fir and red spruce directly by claiming individual trees and indirectly by fragmenting populations, causing erosion, and burning away the upper organic soil layers. The consequences exacted a heavy toll on the spruce-fir ecosystem, and, apparent by both species' conservation status, it has been slow to recover.

The ecosystem's recovery has been significantly delayed by the balsam woolly adelgid (*Adelges piceae*), an invasive insect that was detected in the Southern Appalachians in the 1950s. Thought to have been introduced to the United States around the turn of the twentieth century, likely having arrived on nursery stock imported from Europe, the balsam woolly adelgid has proven a formidable adversary for Fraser fir. In North America, the insect exists only in female form and reproduces by parthenogenesis, in which new individuals are born from unfertilized eggs. Most balsam woolly adelgids are wingless, and their dispersal to uninfested trees occurs during the motile "crawler" stage, the only life phase in which they have legs. Crawlers are dispersed by wind or by animals such as birds, and when they arrive at a new tree, they anchor their mouthparts into fissures in the bark.

Crawlers seem to favor mature firs, particularly those that are 1.5 inches or greater in diameter at breast height. Once attached to the bark, crawlers remain there for the rest of their lives, feeding and laying eggs. As they feed, they secrete threads that coalesce into a white cotton-like blanket, the woolly covering for which they are named. The covering protects the adult female and her eggs, which may number anywhere from 50 to 250. Typically, two generations are produced per year, though occasionally a third or even a partial fourth generation may be produced when warm temperatures endure in normally cool seasons. Symptoms of infestation in firs include gouting (swelling) around buds and shoot nodes and excessive growth of compression wood in the sapwood (xylem) layer, a response that appears to be a sort of allergic reaction to the insect's saliva. Compression wood normally is formed in response to mechanical stress, such as wind. In response to balsam woolly adelgid infestation, however, compression wood grows excessively and eventually interferes with the conduction of water and minerals through the xylem. As a result, the growth of shoots and branches is stunted, and a condition known as flat top, in which the crown of the tree becomes bent, may develop. Infested firs die within about three to nine years.

The first wave of attack by the balsam woolly adelgid in the Southern Appalachians lasted from the time of its detection through much of the 1990s. In the first several decades of that period, scientists reported that the tiny insect had killed more than 80 percent of mature Fraser firs in some areas. Toward the end of it, they were reporting that more than 95 percent had died. Some of that later rise in mortality may have been caused by the combined effects of balsam woolly adelgid infestation, acid rain, and climate change. The latter two factors acted in their own right but together may have significantly increased the fir's susceptibility to the balsam woolly adelgid.

With mature Fraser firs having all but disappeared by the 1990s, the species' survival was left to the small populations of seedlings that remained. Some of those seedlings have since begun to reach reproductive maturity, though their long-term viability remains uncertain. Some of that uncertainty is associated with the disturbance already present in the ecosystem and whether the fir can overcome it. The rest is bound up with signs that the balsam woolly adelgid might be mounting a second assault, supported by the climate change that is looming ever more ominously over the Southern Appalachians.

The presence of balsam woolly adelgid, combined with other forces acting on red spruce and Fraser fir, could steer the ecosystem in one of several different directions.[4] The loss of fir could provide more space for spruce to expand into. Perhaps red spruce would even become the dominant overstory species. Based on research of red spruce–Fraser fir forests in Great Smoky Mountains National Park, that scenario seems fairly unlikely. There, some sites have seen slight increases in red spruce biomass in the overstory, but the spread of spruce into available space has been limited. In fact, about fifteen to twenty years after balsam woolly adelgid had wiped out mature firs, there were noticeable increases in red spruce mortality in the forests studied. The rise in mortality was thought to be associated in part with canopy disturbance caused by the loss of full-grown firs, which left the shallow-rooted red spruce more vulnerable to windthrow. The effects of wind may be increasing in the Southern Appalachians. Or, at least, that is the perception, owing to several recent, especially violent storms in the southeastern and eastern United States, among them Hurricane Hugo (1989), Hurricane Ivan (2004), and Superstorm Sandy (2012). The speculation is that climate change is behind the apparent escalation in storm severity in those regions.

Another scenario that has been proposed is the infiltration of the spruce-fir ecosystem by various deciduous trees, including yellow birch, as well as by certain types of shrubs or invasive plants in the understory. That situation seems plausible, given that canopy gaps created by the loss of Fraser fir may result in increases in the amount of light available to plants on the forest floor. A number of different species whose presence or abundance might normally be limited by overstory shade or other factors might begin to thrive in the gaps. Blackberry and raspberry, for example, could become more abundant. At the same time, herbaceous species and moss, which require damp, dark habitat, could become less abundant. It remains unclear to what extent other species of plants, including deciduous ones, have been moving in to fill the overstory gaps created by the loss of fir. In some areas of the Great Smoky Mountains, young Fraser firs appear to be holding their ground, leveraging their shade tolerance and rapid rate of growth to outcompete other species. Elsewhere, namely in lower elevations, the loss of fir in the overstory may be facilitating the transition from coniferous habitat to open herbaceous or mixed deciduous habitat.

A third scenario entails the complete loss of Fraser fir. The species' situation is tenuous, especially given that it does not retain a soil seed bank, in which seeds would otherwise be able to lie dormant until conditions became more suitable for growth. With the threat of balsam woolly adelgid alone (in the absence of compounding factors such as climate change), the likelihood of the fir's extinction would seem to be fairly low. Young trees have managed to survive thus far, despite habitat disturbance introduced by balsam woolly adelgid. If those young Fraser firs are able to reproduce successfully, then the future of Fraser fir should be secure. Climate change and human activities, however, complicate the picture significantly.

The scenarios are hypothetical, but each warrants some thought, the last one especially. There are numerous pitfalls in attempting to predict the fate of species in the context of climate change. But thinking about those different scenarios helps us prepare for what might lie ahead and gives us a glimpse, if only a fuzzy one, of what might become of our natural heritage in the coming decades. If, for instance, Fraser fir seedlings do not survive over the long term, whether on their own or with the aid of conservation efforts, then we should expect fundamental changes to occur in the Southern Appalachian spruce-fir ecosystem. In the end, despite what researchers are seeing on the ground now, the ecosystem could shift to being dominated by deciduous species, and Fraser fir could disappear.

Red spruce may also be in danger. The existence of multiple threats, ranging from acid rain and climate change, to the native eastern spruce beetle and spruce budworm, to invasive pests like the European sawfly, all of which have affected populations of red spruce, may be facilitating the transition to a deciduous-dominant ecosystem. Those threats exist for red spruce across much of its range, which includes not only the high elevations of the Southern Appalachians but also the lower elevations and sea-level habitats of forests in the northeastern United States and eastern Canada. Thus, even with red spruce's broader distribution, the species' situation is little better than Fraser fir's, and in fact certain of its populations lying to the north of the Southern Appalachians have been devastated. In New Jersey, for example, red spruce is considered to be an endangered species.

Similar to Fraser fir, red spruce is an important species not only ecologically but also in an ethnobotanical sense (ethnobotany is the study of the relationships between people and plants). In the Southern Appalachians, it was known locally as "he-balsam," while Fraser fir was "she-balsam"—names that differentiated the two species based on the absence or presence of resin blisters, respectively. (The use of "balsam" in local reference also reflected Fraser fir's close relationship to balsam fir, *Abies balsamea*.) Native Americans appear to have used red spruce extensively, most likely because it was widespread and because its growth in low elevations in the more northern parts of its range made it readily accessible. The Chippewa, for example, had a recipe for concocting a beverage from red spruce, and they used the tree's fiber in the construction of canoes and its gum for caulk. The Cherokee wove baskets from red spruce fiber and used the tree's wood for various purposes. Today, red spruce continues to be an important timber species, with its wood being used for furniture, musical instruments, and papermaking. In addition, the Northeastern Algonquin and Cherokee used red spruce for the treatment of colds and other respiratory illnesses, suggesting that the species may also be an important resource for substances of medicinal value.

Fraser fir's ethnobotanical significance, by comparison, lies primarily with its importance as a Christmas tree. Under cultivation, it has provided conservationists with a large pool of trees with a history of exposure to the same threats as those experienced by wild firs. Scientists' observations of those farmed trees, combined with information gathered from cultivated study populations, could prove pivotal to conservation efforts.

A major area of interest is whether there might be Fraser firs, cultivated or wild, that possess some degree of natural genetic resistance to balsam woolly adelgid. Seeds from those individuals may be collected and used to propagate resistant seedlings that eventually could be transplanted to areas with suitable habitat. Promising candidates for resistance genes include the few mature Fraser firs that have managed to survive in stands where balsam woolly adelgid infestation has claimed neighboring firs.

In addition, there are species of fir found elsewhere in the world that are tolerant of or resistant to balsam woolly adelgid. Those trees could be hybridized with Fraser fir to produce adelgid-resistant populations of fir in the Southern Appalachians. Scientists have also been working to identify genes within the fir genome that confer resistance to heat, drought, and disease, including root rot caused by the water mold *Phytophthora*. If discovered, those genes could be used to genetically engineer a more robust Fraser fir. That development would have implications for the species' cultivation in lower elevations, such as boosting its production as a Christmas tree and an ornamental, and it could aid conservation in the wild.

In the current period of rapid climatic warming, however, protecting Fraser fir is a difficult and uncertain process. There is, for instance, a fairly high degree of unpredictability concerning how the species will respond to warming. Whether genetic engineering or hybridization with heat-tolerant species is a wise decision for its conservation remains unclear. Those approaches are human interventions that would fundamentally alter the natural state of not only Fraser fir but also the spruce-fir ecosystem. While they would be performed for the benefit of Fraser fir and the ecosystem, because they involve human-driven change in nature, they echo worrisome themes that lie at the root cause of other environmental issues, including climate change. With those issues in mind, it is important that we understand the natural processes that have shaped Fraser fir's existence and supported its survival through history. Those factors are especially significant in light of the fact that for the last several thousand years, with the exception of brief periods like the Little Ice Age (circa 1350–1850), warming has been a near-constant force in North America, very much influencing the continent's natural heritage and the landscape that we see today.

Insight into past climate change and accompanying processes such as species adaptation is necessary to understanding how Fraser fir ended up in those six populations in the Southern Appalachians and how it has managed to hold on there. Insight is also useful for conservation, because it enables officials to choose the most effective strategy or combination of strategies to help ensure the protection of Fraser fir and its ecosystem.

About twenty thousand years ago, the boreal-like forests that covered the central and southern regions of what now make up part of the United States began to change. In the east and southeast, species of cold-adapted conifers, such as spruce and jack pine, began migrating north and west across the North American continent, expanding out of their population centers in and around the Southern and Central Appalachians. Slowly, they trekked over mountains and across valleys, following the retreating edge of the vast Laurentide Ice Sheet, which at its maximum southern extent in eastern North America stretched to within several hundred miles north of Mount Mitchell. By about eight thousand years ago, the size of the ice sheet had become dramatically reduced, and it remained only in the region of modern-day Labrador and Quebec, disappearing from those areas over the subsequent millennium. By that time, too, fir appears to have emerged from glacial refugia at various elevations throughout the Appalachian Mountains. It became abundant in habitat exposed by glacial retreat. In the Southern Appalachians, however, as warming continued, fir and spruce were pruned from lower elevations. Eventually, only those populations that had migrated vertically, out of the lower, warming elevations, were left. The elevations they abandoned became occupied by broad-leaved species, which were better adapted for the warmer climate.

For fir, the discontinuity that was created between populations as they moved vertically in response to climatic warming hindered reproduction and migration and ultimately led to the isolation of southern populations. Those populations were thought to be genetically distinct from balsam fir and the so-called intermediate fir (*Abies balsamea* var. *phanerolepis*), which now populate the northern and central segments of the Appalachians. The three may not be entirely distinct species, however, despite certain genetic and morphological differences.[5] In fact, they appear to still be very close genetically, such that they may be more appropriately considered varieties of balsam fir, which could have implications for Fraser fir's conservation. Their relatedness would mean that the distribution of balsam fir may overlap to an appreciable extent with red spruce, which managed to spread north.

Regardless of species distinctions, because the species that we currently know as Fraser fir lingered in the southern mountains, whereas red spruce spread beyond them, Fraser fir can be considered a glacial relict, a remnant of that once-vast boreal-type forest that thrived south of the Laurentide. The Laurentide, which by some estimates measured a massive 2 miles in thickness, was a hulking slab of ice that drew moisture from the atmosphere. Thus, the climate in which ancient boreal forests and boreal-type flora thrived likely was both cold and dry, presumably something like the tundra biome that now exists across much of Alaska and northern Canada.[6] As the ice retreated, cold-adapted boreal species followed it in pursuit of the cool climate that lingered near its edge. The boundaries between boreal plant communities marching north and the deciduous communities infiltrating in the south corresponded roughly to the locations of major air masses. In the boreal zone, for example, there would have been a cool Arctic air mass, favorable for boreal vegetation, whereas in the more southerly deciduous zone, a warmer Pacific air mass would have dominated. The delineations between those air masses appear to have been influenced by the presence or absence of ice and by certain other factors, such as topographical features in the land. At its southernmost point prior to the Laurentide's retreat, the cool Arctic air mass extended into modern-day Georgia and Alabama. As it was drawn northward with the retreating ice, warm air eventually penetrated into higher latitudes. Those changes likely also brought more pronounced seasonality to the southern and central latitudes of North America, in turn influencing the life cycles of plants, including the timing of bud break and flowering.

In the lower parts of the Southern Appalachians, populations of boreal trees and other boreal-like flora that did not follow in the wake of the retreating ice underwent extinction. Presumably they went extinct because they could not adapt to cope with the warming climate or, in the case of populations of cool-climate coniferous species, because red spruce and fir outcompeted them for habitat in the high elevations. Thus, despite their best efforts at adaptation and migration, those populations simply were overcome. Certain populations of spruce and fir, on the other hand, fortuitously found themselves in pockets of habitat suited for their survival, most likely the few places left in which local climate had remained relatively cool and moist.

The need for a cool, moist climate, as opposed to a cool, dry one, might also explain why the fir that we now know as Fraser fir migrated vertically rather than northward, or at least why it was able to persist in the southern mountains while other cold-adapted conifers died off or moved away. Fraser fir's overall response to a warming climate may have been similar to that of its coniferous cousins that ended up in the north. It migrated, but in a different dimension, and at the same time, it likely was becoming better adapted to its new environment, adjusting to the relatively rocky, shallow soils found on the high slopes, for example. Based on these assumptions, we could conclude that other species of high-elevation forest trees cope with changing climatic conditions through a combination of strategies, such as migrating to suitable habitat while also investing energy to adapt to the various new conditions they encounter along their journey.

One reason why it is difficult to anticipate how some populations of plants will respond to climate change is the fact that many populations, including those of relict species, cannot migrate. An increase of 5.4°F could raise current ecological boundaries between biological communities 1,600 feet, according to some estimates.[7] That would push Fraser fir clear off the mountaintops of its habitat. There are no higher peaks to support vertical movement. Nor do there seem to be high peaks located within reach of the species' natural migration distance. In the end, it may have no choice but to endure in its current location, which would mean that in order for it to survive, assuming that humankind does not intervene, it would need to adapt to the warmer, drier conditions that could overtake its habitat.

Unfortunately, the swiftness in onset of recent changes in climate does not bode well for those transitions, as it has left little time for plants to adapt or migrate. Estimates suggest that the pace of migration for some trees is slow, around 330 feet per year. Some tree species are able to migrate faster, but not by much. For example, beech (*Fagus*) and maple (*Acer*) may move distances of 560 to 700 feet per year. But models based on a doubling in levels of atmospheric carbon dioxide over the next century, which would correspond roughly to an increase of about 5.5°F by the year 2100, predict that boreal trees would need to migrate ten times farther each year in order to keep up with climate change–induced geographical shifts in their habitats. According to the Intergovernmental Panel on Climate Change

(IPCC), by the end of the twenty-first century, global surface temperatures could be anywhere from 3.2°F to 7.2°F higher than they were at the beginning of it.[8] It seems almost certain, then, that at least some of North America's native plant species will undergo redistribution and that they will need to effect that process quickly.

Three Plants and Their Animals

Each May, the jack pine ecosystem of the northern Lower Peninsula of Michigan witnesses the unfolding of an amazing relationship between plant and animal. At that point in the year, Kirtland's warblers (*Setophaga kirtlandii*) arrive and set up their summer homes in jack pine (*Pinus banksiana*) ecosystems. The birds fly up from the Bahamas and surrounding islands for the warm season, traveling 1,200 miles to reach just those trees in Michigan. Although breeding Kirtland's warblers have been observed in Michigan's Upper Peninsula, as well as in Wisconsin and Ontario, Canada, it is among the jack pine forests of the northern Lower Peninsula where the vast majority breed, specifically in the region extending south from Presque Isle County to Clare County and west to Wexford County.

There is something special about the jack pine ecosystem in the northern Lower Peninsula. Jack pine is an otherwise widespread species. In Canada it is one of the country's northernmost pine species, being found from Nova Scotia to northeastern British Columbia. In the United States it occurs in the northeastern and northern Midwest regions, the latter area encompassing Michigan, northwestern Indiana, northeastern Illinois, Wisconsin, and Minnesota.[1]

Across its range, jack pine occurs in different forest and barrens ecosystems, and it associates with a wide range of other tree and plant species. Its plant associates typically differ from one region to the next, as does its dominance within its forest communities. In the forests of the northern Lower Peninsula, for example, a common tree associate is northern pin oak (*Quercus ellipsoidalis*). In forest areas elsewhere in its range, jack pine may grow

40

alongside aspen (*Populus* spp.) and birch (*Betula* spp.), red pine (*Pinus resinosa*) and white pine (*P. strobus*), oak (*Quercus* spp.) and hickory (*Carya* spp.), or spruce (*Picea* spp.) and fir (*Abies* spp.).

In its forest communities, jack pine often is the characteristic species, but sometimes it shares dominance with other trees, notably aspen, red pine, and paper birch (*Betula papyrifera*). In forest ecosystems, it also associates with certain undergrowth and ground-layer species. In the jack pine–oak–*Arctostaphylos* ecosystem at Voyageurs National Park in Minnesota, for example, jack pine has been found to associate with a variety of understory species, including common juniper (*Juniperus communis*), cup lichens (*Cladonia* spp.), eastern teaberry (*Gaultheria procumbens*), kinnikinnick (*Arctostaphylos uva-ursi*), and smooth sumac (*Rhus glabra*).[2]

In jack pine barrens, which are known primarily from the Great Lakes region, including areas in Michigan, northern Wisconsin, and northern Minnesota, jack pine forms much of the overstory. Other conifers, however, particularly red pine, may also occur, occasionally even forming canopies over jack pine. Jack pine barrens also are home to other types of trees, including aspen, black cherry (*Prunus serotina*), and northern pin oak. Often, those species occur as young or short trees. Beneath them sit low-growing plants such as lowbush blueberry (*Vaccinium angustifolium*), prairie willow (*Salix humilis*), sand cherry (*Prunus pumila*), and sweet fern (*Comptonia peregrina*). Numerous herbaceous species, such as birdfoot violet (*Viola pedata*), little bluestem (*Schizachyrium scoparium*), Pennsylvania sedge (*Carex pensylvanica*), porcupinegrass (*Stipa spartea*), poverty oatgrass (*Danthonia spicata*), and sky blue aster (*Aster oolentangiensis*), can be found in the ground cover. Typically, shrubby species and other ground-cover plants grow to no more than 6 or 12 inches in height.

It is, in part, the assemblage of plant species to be found in the jack pine ecosystem of Michigan's northern Lower Peninsula that attracts Kirtland's warblers. Landscape-level diversity is a characteristic of the species' preferred habitat, with various low-growing shrubs serving as sources of food, for example. But equally important for Kirtland's warblers is that the habitat area is large, covered by a mosaic of open areas interspersed with dense patches of trees and underlain by permeable soils. Those characteristics provide space for the establishment of territories, cover for nests and fledglings, and drainage for nests, which sit on the ground.

Female warblers and chicks spend much of their time on the ground in their nests, and the dry, sandy soils that are characteristic of the jack pine ecosystem provide the perfect foundation for ground nests. In addition, surrounding bits of sedge, shrub leaves, pine needles, and other plant materials are used in nest construction. Whereas the soil characteristics essentially are more or less stable in jack pine ecosystems, the ability of the ecosystem to provide adequate nest cover and food for foraging birds is a function of its age. Jack pine stands that are used by Kirtland's warblers range from six to about twenty-two years of age. At six years, young jack pines are about 5 or 6 feet tall, and their branches fill in along the trunk, extending almost all the way down to ground level. Various low-growing plants are dense beneath the trees, helping to provide cover for the nests. Young chicks, which leave their nests just nine days after hatching, take to the undergrowth and low branches, finding shelter and safety there.

By twenty years of age, jack pines are between about 13 and 20 feet in height, and their lowest branches may be a few feet off the ground. Their branches expand as they age, casting shade over low branches and much of the ground cover and in turn preventing the growth of branches low enough to touch the ground cover, as well as most ground-cover species. As a result, the amount of open, shaded space beneath the trees increases. It is around the time that low branches begin to disappear and the ground beneath the trees becomes more exposed that the warbler begins to abandon a jack pine stand. That process appears to take place anytime after a stand reaches fourteen years of age. The density of jack pines seems to be a factor, too. Areas where young trees are dense and where about 25 percent of the habitat is open are especially attractive to Kirtland's warblers.

In the past, the warbler's habitat specifications were determined largely by fire, which was the primary disturbance in jack pine ecosystems. Fire aids the regeneration of jack pine in two ways: through the delayed release of seeds from its serotinous cones and through the creation of habitat. Cones are most numerous in the crown of the jack pine, so fires that involve the crowns of trees are the most productive for the species in terms of cone opening and seed release. The winged seeds, sitting loose in the cones or having fallen out onto the charred ground, are dispersed by the wind. Suitable habitat, in which the seeds can germinate and produce seedlings successfully, is formed by a burn as it exposes the mineral soil seedbed and kills off mature jack pines and other plants.

Jack pine is a pioneer species in burned sites. After its seeds are freed from cones and dispersed onto the seedbed, germination ensues. Once established, seedlings develop quickly, maturing and producing flowers as young as five years when growing in open stands. Jack pine, in fact, is one of the fastest-developing and fastest-growing pioneer tree species in its ecosystem. It is shade intolerant and must outcompete other trees for sunlight to become established. Eventually, its growth rate slows, however, resulting in a relatively small- to medium-sized tree. Jack pine is at its shortest and scrubbiest in dry sandy or pebbly soils, which are relatively infertile. In dry loamy soils, it grows taller, typically peaking at heights of about 55 to 65 feet and achieving a diameter of about 1 foot or less. Jack pine stands, in the absence of fire, appear to be able to thrive until about sixty to eighty years of age, at which point the trees grow very slowly. The oldest jack pines likely are to be found on relatively infertile sites, where scrubby individuals may persist naturally without fire and without succession. On most other sites, however, such as those with dry, loamy soils, and particularly when fire is suppressed, jack pine eventually is succeeded by other species.

Historically, fires probably were fairly frequent in jack pine ecosystems. That assumption is based in part on jack pine's serotinous cones, which are sealed by a resin that secures the seeds inside. Serotinous cones are opened by exposure to high temperatures, which often come in the form of fire. Cone serotiny in jack pine presumably has been with the species for some time, perhaps since shortly after its arrival in its current range, when glacial retreat enabled it to escape warming temperatures in its southern refugia in the Appalachia region. Fire likely played a major role in the historical development of forests in the Great Lakes region. Data from various studies have suggested that burns occurred, on average, about every four to thirty-six years in forests where jack pine, red pine, white pine, aspen, or northern pin oak were common. Likewise, estimates made in the early 1980s of the frequency of fire dating back to 1839 in jack pine ecosystems near Mack Lake in Michigan's Oscoda County suggest that fire occurred there at regular ten- to forty-year intervals.[3] The assumption that fire was frequent historically in the region where jack pine is found is also based on charcoal evidence. In the 1970s, for example, researchers detected elevated charcoal levels in lake silt from the Boundary Waters Canoe Area Wilderness in Minnesota. The samples were dated to between six thousand and nine thousand years ago and were linked to a preponderance of jack pine and red pine.

How all those fires were ignited is not entirely clear, though drought likely was a common facilitator, enabling some fires to be ignited by lightning strikes. Other fires came at the hands of Native Americans, intentionally or unintentionally. The decline of fire frequency since the mid-eighteenth century corresponded somewhat with the penetration of early European pioneers into and the beginning of the extirpation of Native Americans from the Great Lakes region. Evidence for the decline of fire is apparent in the charcoal levels in Boundary Waters lake silt, which were found to be lower in sediments dated to the post-mid-eighteenth century compared with sediments from before that period. The most marked reductions in fire in Midwestern states occurred following the implementation of fire-suppression policies in the 1920s. Fire was suppressed primarily to prevent losses to property and human life. But as natural fire-adapted habitats such as prairie, savanna, and woodland areas were converted to agriculture, resulting in changes in soil chemistry and ignitable plant fuel, the opportunity for fire was reduced, too. Forested lands that were left intact advanced in succession, resulting in the elimination of trees regenerated by fire.

How large a role fire suppression played in the decline of Kirtland's warbler is unclear, but over the course of a few decades in the twentieth century, the warbler became one of North America's rarest species of songbird. By 1987 just 167 singing males were left. By that time, too, availability of suitable breeding habitat had decreased significantly. In the early 1980s, just under 25,000 acres of suitable, early-successional jack pine habitat—the kind of habitat typically created by wildfire—remained for breeding Kirtland's warblers. The implication of habitat loss in the bird's decline was strengthened by the subsequent rebound in the 1990s and early 2000s of both breeding habitat and populations of Kirtland's warblers.

More than habitat loss was at play, however. A factor that was especially detrimental for Kirtland's warbler was nest parasitism by brown-headed cowbirds (*Molothrus ater*). The conversion of forests into agricultural fields in the late nineteenth and early twentieth centuries facilitated the expansion of the range of the brown-headed cowbird from the Great Plains into Michigan. By the 1940s and 1950s, cowbirds were having major impacts on warblers, parasitizing an estimated 55 percent of warbler nests. Parasitism was associated with reduced reproductive success for Kirtland's warblers, primarily because it led to reductions in warbler clutch size and fledgling production. In the 1970s, following the introduction of measures to control cowbirds, which consisted mainly of annual trapping of the birds, parasitism rates dropped. Still today, however, brown-headed cowbirds return in massive numbers each year, making their control a major part of habitat management for Kirtland's warbler.

In total, some 219,000 acres of jack pine ecosystem in Michigan have been designated for Kirtland's warbler management. Each year, several thousands of these acres are developed into breeding habitat, and several tens of thousands of acres across the managed area are suitable for breeding. Although prescribed burning is desirable for jack pine ecosystems, the fires needed to regenerate jack pine pose great risks to property and human safety in some areas. As a result, relatively few acres of jack pine habitat in

Michigan are burned intentionally. Rather, in most cases, management for Kirtland's warbler has centered on mimicking the effects of fire as much as possible. In some areas, the ecosystem has been managed successfully through clear-cutting and replanting of jack pine. Researchers with the US Forest Service and the Michigan Department of Natural Resources have found that the planting of trees at relatively high density and in an opposing wave pattern, which increases the diversity of open spaces, is especially beneficial for Kirtland's warbler. About 25 percent of the regenerated area is left open.[4] The management strategies have helped Kirtland's warbler recover. In 2012 a record 2,090 singing males were counted. The numbers are encouraging, but maintaining them depends on continued ecosystem management.

While secure in Michigan, the future of jack pine elsewhere is uncertain. It is considered a rare species in Indiana and New York, a threatened species in New Hampshire and Vermont, and an endangered species in Illinois, all of which are areas that fall within its current range. Since the early 1980s in Wisconsin, there have been significant declines in jack pines of nearly all age classes, with a 74 percent drop in average annual net growth.[5] The volume of jack pine harvested in Wisconsin gradually began to outpace its growth rate, with the latter hindered significantly by high mortality. By 2012 the removal rate was twice that of jack pine's rate of growth, and although the reasons for increased mortality and slowed growth were unclear, disease and pests, such as jack pine budworm (*Choristoneura pinus*), may have been contributing factors. In Canada jack pine is threatened by mountain pine beetle infestation.[6] The beetle recently expanded its range across the Rocky Mountains and into north-central Alberta, where forests shift in composition from lodgepole pine to jack pine. Patches of jack pine and jack pine–lodgepole pine hybrids run eastward across Canada, creating a corridor that could facilitate the migration of mountain pine beetle east to previously unaffected forests.

As those declines and threats have emerged, however, Kirtland's warblers have been increasingly sighted in jack pine ecosystems in Wisconsin and Ontario. Before 1995 the warbler was rarely seen in North America outside Michigan's northern Lower Peninsula, so its presence in other areas is promising. It is an indication that suitable breeding habitat exists beyond Michigan and that jack pine forests in those other areas are healthy. And so it is that each year, when Kirtland's warblers complete their remarkable journey home to their birthplaces, they have another opportunity to perpetuate their species' existence, and we have another opportunity to celebrate jack pine.

Many animals have an instinctual drive to return to their place of origin, their home, as we tend to think of it. An animal's birthplace often has deep ancestral roots. It is a place where generations of a particular herd or flock or group have been born and have returned, faithfully, year after year. Some animals, many perhaps, are now unable to trace their ancestors' paths. Humans and human development have gotten in the way, disrupting migratory paths and extirpating animals from large swaths of their homelands. The consequences to the animals affected have been great, and so, too, have been the consequences to the plant species that depend on those animals.

We rarely hear about the latter, but some species of plants have suffered greatly as a result of the extinction or extirpation of their animal associates. Among the more extraordinary examples is running buffalo clover (*Trifolium stoloniferum*), which historically depended on the presence of American bison (*Bison bison*) in primarily three geographic regions: the Appalachia of Kentucky, Ohio, and West Virginia; the area that is now Kentucky's Bluegrass region; and the Ozarks. About a century and a half following the near extinction of American bison in those areas, running buffalo clover had declined very low. In some places, it had died out completely.

The story of what came to pass for both species is a part of the history of those three geographic regions. European settlers began to populate the Appalachia regions of Ohio and Kentucky in the eighteenth century. But long before the arrival of Europeans, Native Americans had laid claim to the land. Indigenous groups of people had occupied the region for at least ten thousand years, though which tribes' ancestors were where is uncertain. By the time Europeans arrived there, however, the land had been clearly divided among different tribes. The Shawnee were in the north, occupying much of Ohio and Kentucky; the Yuchi and Cherokee were to the southeast; and the Chickasaw were to the southwest. All were sustained, to varying degrees, by hunting, agriculture, and trade.

Not surprisingly, there were numerous conflicts between the new settlers and the Native Americans, often deadly to the former, at first. But strength was in numbers, and eventually, as more and more Europeans moved in, the tribes of Native Americans were forced out. The new arrivals did their best to tame the wild land, converting what areas they could for agriculture, and they seeded it with types of grass suited for livestock grazing. Perhaps most well known among those introduced species were bluegrass (*Poa pratensis*) and timothy grass (*Phleum pratense*).

Europeans, similar to Native Americans, also practiced hunting. But unlike the bows and arrows, blowguns, and spears in the Native American hunting repertoire, Europeans had rifles. And they used them at will, hunting animals occasionally for food but mostly for profit or sport. Populations of large game animals, including deer, elk, and bison, crashed. By 1800, eight years after Kentucky achieved statehood, American bison had been hunted so intensely that they were nearly locally extinct. The timing of the relentless persecution of bison in the Appalachia region corresponded with the beginning of running buffalo clover's decline.

Running buffalo clover is a perennial, low-growing clover that favors limestone-rich soils. It bears leaves with three leaflets and produces a single, spiked, roughly inch-wide white flower, sometimes with purple accents. Flowers typically appear between mid-April and late June. The flower head is supported on a stem that ranges from about 4 to 12 inches in height, and beneath it extends a pair of large, opposite aerial leaves, a distinguishing characteristic of the species.[7]

The type of setting where running buffalo clover occurs varies. It has been found in mesic woodlands and grazed woodlands, on floodplains, and in steep ravines, as well as along game, foot, and vehicle trails and in other areas disturbed by grazing or trampling. It also has been found on sandbars by seasonal streams or at points where trails cross or run alongside the water and on lawns of historic homes and cemeteries managed with periodic mowing. American elm (*Ulmus americana*), black walnut (*Juglans nigra*), boxelder (*Acer negundo*), sugar maple (*A. saccharum*), and white ash (*Fraxinus americana*) typically are found in the surrounding overstory. Herbaceous plants that grow near running buffalo clover include various grasses (*Poa* spp.) and sedges (*Carex* spp.), common chickweed (*Stellaria media*), hog peanut (*Amphicarpaea bracteata*), white clover (*Trifolium repens*), and white snakeroot (*Ageratina altissima*), among others.

The full extent of the natural range of running buffalo clover includes northern Arkansas, southern Illinois, central and southern Indiana, eastern Kansas, central Kentucky, southern Missouri, central and southern Ohio, and central and northern West Virginia. The species was thought to be extinct throughout those areas until 1985, when a pair of populations was identified in West Virginia. They were found in a mountainous area, one along the edge of a dirt road and the other in a lawn beside a gravel road. Since then, naturally occurring populations have been discovered across parts of its range, with the exception of Arkansas, Illinois, and Kansas, where it appears to have been extirpated. Running buffalo clover has been on the US List of Endangered and Threatened Wildlife and Plants since 1987. It is also listed as endangered separately in Indiana, Missouri, and Ohio. In Kentucky, which has the greatest number of known populations (about forty-six extant in 2007), it is listed as a threatened species.[8]

Figure 4. The flower of running buffalo clover (*Trifolium stoloniferum*). (Credit: Kara Rogers)

Running buffalo clover's runners, also known as stolons, creep along the ground, sometimes becoming buried beneath the soil. Nodes along its length send down adventitious roots, and buds give rise to new plants, each the same genetically. In other stoloniferous plants, stolons facilitate the sharing of certain nutrients, such as nitrogen, between individual plants. In running buffalo clover, that type of nutrient sharing may help plants make use of resources available in disturbed habitats. Stolons eventually break up, and each rooted individual becomes a separate daughter plant, capable of sending out its own runners.

While regeneration by stolons enables the rapid propagation of plants, a significant downside of that process is the restriction that it places on genetic diversity. All the plants on a given stolon are clones, and with uniformity across their genes, their ability to respond and adapt to environmental change presumably is limited. Often in species that propagate by stolons, the greatest levels of genetic diversity are seen between populations rather than between individual plants. That is the case with running buffalo clover, in which small and large populations contribute equally to gene diversity.

In addition to the species' clonal nature, other forces may also be capping heterogeneity at the genetic level in running buffalo clover. Examples include the apparent absence of gene transfer from one population to the next and breeding between related plants. Individuals can self-fertilize, too. As populations of running buffalo clover began to decline, they may have come to depend more heavily on their self-compatibility than on outcrossing, which requires the aid of pollinators such as bees. The result would have been a reduction in fertility and survival in subsequent generations, which would have sent the species into decline. It is unclear, however, whether inbreeding depression had a hand in the decline of running buffalo clover, given that self-fertilized individuals have been known to thrive. Still, it is possible that the combination of those characteristics might explain at least some part of the struggle that running buffalo clover has faced since American bison disappeared from the clover's habitat.

Running buffalo clover is an example of the magnificent and unexpected nature of plant–animal relationships. In its native habitat, it lived with bison. It associated with the massive animals, growing alongside their travelways (traces) and in the savanna woodlands the bison helped maintain. A major reason for that association appears to have been that the American bison was a destructive animal. It was vast in number and truly enormous in size, standing between 5 and 6.5 feet at the shoulder and weighing from 900 to 2,200 pounds. Everywhere it went, it beat up vegetation and tore up soil. And amazingly, despite its comparatively diminutive size and fragility, running buffalo clover depended on the bison's environmental abuse to thrive. Bison ate the clover, helping to scarify and disperse its seeds. They also packed down soil in their wallows and roughed soil up along their traces, activities that limited the growth of trees. The result was a sparsely wooded habitat, a savanna woodland, that offered the ideal mix of sun and shade for running buffalo clover.

Since the bison's disappearance from the Appalachia region, and perhaps beginning in parallel with it, other factors, too, have contributed to the loss of running buffalo clover. The loss of habitat and habitat change, mainly through altered land management practices that facilitated ecological succession in formerly open woodlands, are factors that likely came into play when European settlers moved in. Those factors grew in severity as the region became increasingly developed. Another issue that probably has been significant as well is the introduction of nonnative species. Garlic mustard (*Alliaria petiolata*), Nepalese browntop (*Microstegium vimineum*), Japanese honeysuckle (*Lonicera japonica*), and white clover, which is native to Europe, have been especially problematic for running buffalo clover. Those species soak up nutrients, space, water, and sunlight. Introduced clovers also are not shy about sharing the diseases and pests they carry.

Overcoming the various threats is vital to running buffalo clover's restoration. The challenges are significant, but fortunately some of the solutions are relatively uncomplicated. Regular surveying of populations helps researchers keep tabs on the number and health of individuals. Control for invasive species, strategies for limiting forest succession, and the implementation of grazing or mowing regimes help maintain the existence of running buffalo clover habitat. Public outreach and education, particularly for those who live in areas with ideal habitat for running buffalo clover, raise awareness and encourage people's involvement in conservation efforts. Together, these approaches could bring significant benefits for the species' recovery.

Success has already been seen, too. For example, at Wayne National Forest in southeastern Ohio, where running buffalo clover was found in 2005, botanists counted increasing numbers of rooted and flowering plants in surveys carried out in the following years. Habitat management strategies employed at the site included the removal of nonnative species such as Nepalese browntop and attempts to decrease all-terrain vehicle use in the area. US Fish and Wildlife Service biologists were also working with interested private landowners on the removal of nonnative plant species that were detrimental to the clover. That type of collaboration and stewardship is essential to the restoration of native plant species on public and private lands.

Another illustration of the vital connections that exist between native plants and animals in North America takes us to the American Southwest. A hesitant song, starting "hu-wee chu-wee che-weet," rises up from a sun-bathed canyon slope. The song quakes. It is not as smooth or as pretty as one might expect from a practiced songbird. But it is fitting for the scene, a patch of Sonoran Desert in southeastern Arizona in winter, where the cover of vegetation is interrupted here and there by drab gray rocks of all sizes and resting at various angles. The land is visually jarring, and its beauty manifests in unusual ways.

The performer is difficult to spot, but he sings long enough to give away his location to those patiently awaiting an opportunity for a glimpse. And once seen, he seems clear as day. His dull gray plumage, inconspicuous almost every place else in the bleached desert, stands out subtly against the backdrop of succulent green leaflets, purple fruits, and red twigs of an elephant tree (*Bursera microphylla*).

The bird is a gray vireo (*Vireo vicinior*), a type of North American songbird. Its most distinguishing characteristics are a thin white eye ring, a faint white wing bar, and a longish tail, which it flicks up and down at regular intervals. Overall, it measures about 5.5 inches in length and weighs about half an ounce. It is warbler-like in appearance, except for its thick bill.

During the breeding season, the gray vireo's range runs across parts of Arizona, southern California, Colorado, New Mexico, Utah, and western Texas and extends south into northwestern Mexico, including Baja California. Its winter range shifts south and west and generally overlaps with the Sonoran Desert, which covers southeastern California and southwestern Arizona, as well as the western portion of the state of Sonora in northwestern Mexico and most of Baja California. Gray vireos have also been known to winter in southwestern Texas, particularly in the Chisos Mountains in Big Bend National Park. Individuals appear to return to the same winter territory every year.

In the breeding season, gray vireos forage primarily for arthropods such as caterpillars, cicadas, and grasshoppers. They pluck many of these creatures off the leaves, twigs, and branches of small trees in thickets, though they can also catch them on the wing. The birds sometimes stalk and pounce on their prey as well. In their wintering season, from about late September to early April, gray vireos continue to forage for insects. A couple months into the season, however, migrants in the Sonoran Desert eat primarily fruit, concentrating on elephant trees. They appear to be able to digest only the pulp of the fruit, and so it is thought that they regurgitate the seeds, presumably in viable condition.

It is not an altogether pleasant thought to have one's progeny dispersed by regurgitation, but elephant trees may depend on the process, or at least that is what research has suggested.[9] American naturalist John M. Bates was one of the first to study the relationship between the gray vireo and its winter habitat. In his thesis work, conducted in the 1980s, Bates concluded that the gray vireo served as the primary means of seed dispersal for the elephant tree. As the tree's fruit ripened in the winter season, it appeared to make up a major portion of the bird's diet. Bates's observations further suggested that, within the limited extent of its winter range, the gray vireo appeared to consume only the fruit of the elephant tree. The discovery of seeds from desert mistletoe (*Phoradendron californicum*) in gray vireo droppings indicated that the birds supplement their diets with fruit from other plants, but maybe only occasionally. Likewise, while ash-throated flycatchers were observed feeding on elephant tree fruit, gray vireos seemed to be the dominant consumers. The findings suggested that a mutualistic relationship might exist between the gray vireo and the elephant tree, in which the vireo helps the tree by spreading its seeds, and the tree helps the vireo by supplying it with food through the winter.

It appears that the gray vireo has declined across parts of its range, though whether that really is the case and the true extent of its contraction are unknown. In southeastern California, for instance, it appears to have become relatively rare, but data are insufficient to identify a strong downward trend. Some people think that its numbers have dropped in that state in part because of nest parasitism by brown-headed cowbirds. The actual impact of the brown-headed cowbird on gray vireo populations, however, remains unclear. The factors that point toward the cowbird's possible involvement include the modification and fragmentation of gray vireo habitat, which may have created habitat edges ideally suited for brown-headed cowbirds. The gray vireo's habitat is somewhat naturally fragmented, too, which means that when the cowbird arrived in southern California, it probably had little trouble finding suitable places to take up residence.

Habitat loss probably is related to the gray vireo's presumed decline. Gray vireos make use of chaparral, thorn scrub, and sagebrush habitats, as well as pinyon-juniper, oak-juniper, and juniper-cholla habitats. In California, pinyon-juniper woodlands have undergone extensive modification in some areas, possibly to the detriment of gray vireos. A reduction in gray vireo populations in the species' winter range in California could theoretically contribute to the endangerment of elephant trees there. The same would be true in other parts of the Sonoran Desert where the bird may not be as abundant as it once was in winter. Alternatively, or as a consequence, reductions in elephant tree populations could be inciting declines in gray vireo populations.

The elephant tree is relatively sparse across its disjointed range. Its favored habitat is in the washes and on the rocky plains and slopes in the hot, dry deserts of southwestern New Mexico, southern Arizona, and southeastern

California. In California it is best known from the Anza–Borrego Desert State Park, which lies between San Diego and the Salton Sea. In Arizona it is found in several counties, including Maricopa, Pima, Pinal, and Yuma. It often is found on mountain slopes in those regions, such as on the slopes of the Gila, Tinajas Altas, and Tule Mountains in Yuma County and South Mountain in Phoenix (Maricopa County). It grows at elevations up to about 2,500 feet. The elephant tree is also found in Mexico, specifically in the states of Baja California and Baja California Sur (the lower part of the peninsula), Sinaloa (on the country's northwestern coast), Sonora, and Zacatecas (embedded in the country's north-central region).

A specimen from the region of Cabo San Lucas, on the Baja peninsula, supposedly inspired American botanist Asa Gray to give the species its formal name, which was published in 1861.[10] The species was rumored to exist in the United States in the early twentieth century, but it was not until 1937 that the elephant tree was convincingly rediscovered in several canyons in California's Anza–Borrego Desert.

The elephant tree's common name comes from the shape of its trunk, which tapers vertically in its upper half and is disproportionately thick compared to the tree's overall height, which ranges from about 5 to 25 feet. The unusual thickness of the trunk is typical of burseras, but in the elephant tree, the trunk contains tissues that are able to swell with water, augmenting its storage. The elephant tree can store enough water to see it through a year without rain. There is only one other bursera found in the region that is similarly succulent, making it suited for life in an arid climate. That species is the red elephant tree (*Bursera hindsiana*). The burseras (family Burseraceae) are otherwise tropical and subtropical plants, and relative to the elephant tree, even the next-northernmost member, the red elephant tree, seems to prefer a somewhat moist and breezy environment. It favors the maritime climate of Baja California, where it is common, over the dry desert of Sonora, where it hugs the western coast. Still, both the elephant and red elephant trees are the only known sarcocaulescent, or fleshy-stemmed, burseras, and each may represent a transition in bursera evolution as the group migrates northward into the American Southwest.

The elephant tree's Latin species name, *microphylla*, means "small-leaved," which describes its numerous smooth, pinnate leaflets (ranging from seven to more than thirty on a single leafstalk). Some of the leaflets measure a mere quarter of an inch in length and less than a tenth of an inch in width. The possession of such tiny leaflets means less water loss, a beneficial trait in the desert, where the ability to conserve water, along with the capacity to store it, are of utmost importance. The lack of photosynthetic leaf material in the canopy is made up for by the tree's thin, light-colored bark, which allows sunlight to penetrate into the photosynthetic tissues beneath. The bark becomes increasingly opaque as the tree ages, but it also is exfoliative, such that in dry periods, darker layers peel away in thin sheets. That exfoliating quality may also help prevent certain types of lichens and other growths from forming on the bark and blocking the absorption of sunlight.

The leaves of the elephant tree grow on reddish twigs, and while the species is deciduous, the leaves may remain on the branches year-round for two or three years. Conditions of drought or frost can cause the leaves to fall off, in which case they typically begin their growth again in summer during or immediately after flowering. The tree spits out a stream of aromatic but foul-tasting sap when a turgid stem is broken, which may help protect its leaves from herbivores. In the Sonoran Desert, flowers of the elephant tree usually bloom in June or July. The flowers themselves keep a low profile, being tiny, and occur singly or in small clusters. Fruit set occurs in September. The drupe-like fruits measure about one-quarter to one-third of an inch in size, mature to blue to purple in color, and contain seeds with a thin red aril (exterior covering). The elephant tree also produces a bright red sap, which bleeds from cut wood and roots.

Bates observed fruit lingering on elephant trees into April, and his bomb calorimetry experiments revealed interesting details about the species' fruit. The elephant tree produces a few thousand more fruits than the red elephant tree, though its fruits ripen at different times, which limits the size of the crop that can be taken at any one time. In addition, while the weight of the elephant tree's aril is less than that of the red elephant tree, the former is more calorie rich, coming in at 32.8 kilojoules per gram, to the red tree's 27.4. Bates noted that the elephant tree's caloric value was greater than that recorded for twenty other species of plants (all in Illinois) known to supply fruit to migrating birds. He proposed that the lingering presence of fruit on the elephant tree in spring may be a vital source of energy for birds preparing to migrate to their breeding grounds.

There are several species of fruit-bearing trees and shrubs that share habitat with the elephant tree, and so it could be that the elephant tree is adapted to attract specifically the gray vireo to increase its chances of seed dispersal. For example, the extent to which the gray vireo can open its beak (its gape) is matched perfectly with the size of the elephant tree's fruit. In addition, the elephant tree's energy-rich fruit, its long season of fruit production, and its limited crop size (Bates reported only one or two ripe berries at any given time on a single tree) all point toward a possible coevolutionary relationship with a fruit-eating bird. Whether such a relationship really exists with the gray vireo, however, is unclear.

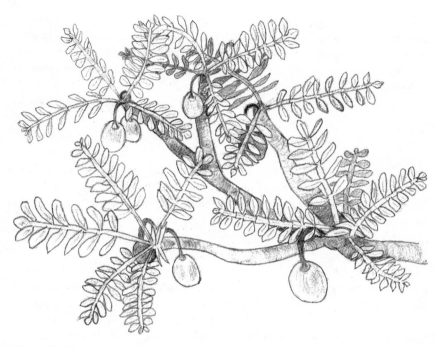

Figure 5. The leaves and fruit of the elephant tree (*Bursera microphylla*). (Credit: Kara Rogers)

Interestingly, there are patches of elephant trees that seem ideal for winter foraging by gray vireos but that have never or have only rarely been seen to be visited by the bird. It is an inconsistency for which a viable explanation is wanting. Perhaps those trees are in trouble, and it is just a matter of time before those populations become extinct. However, the situation raises questions about the exclusivity of the relationship between the elephant tree and the gray vireo. For example, along the coast of the Gulf of California in Sonora, chipmunks have been rumored to eat the tree's fruit, suggesting that maybe they, too, help to disperse the elephant tree's seeds. Thus, in the vireo's absence, and perhaps in its presence, other animals might participate in seed dispersal. Whether the elephant tree can get by with those species as primary seed dispersers if the gray vireo declines substantially remains to be seen.

The elephant tree also shares a unique relationship with humankind. Its family is sometimes called the incense tree family or the torchwood family, because each member's leaves and sap are highly aromatic, similar to their Old World relatives, frankincense (*Boswellia* spp.) and myrrh (*Commiphora* spp.). Also similar to those counterparts, which are probably best known for their depictions in religious texts, the elephant tree has long been used by local peoples, who sometimes refer to the species as *torote*. Copal, the name given to the resin of various burseras (and other trees), was used in religious rituals in the pre-Columbian era and later as a varnish (some indigenous peoples also used the word "copal" to refer to the tree itself). The tree's sap was also used as a dye. The Seri, a group of indigenous people who historically inhabited northwestern Mexico and islands in the Gulf of California, used the resin of the elephant tree as a caulk to plug holes and cracks in pottery and boats. They used the twigs to make tea and employed various parts of the tree medicinally to treat conditions such as cuts, headaches, skin diseases, and sexually transmitted afflictions. They also used the elephant tree for fuel and as a fumigant to smoke bees out of their hives when collecting honey. Extracts prepared from the tree's branches may still be used today for the treatment of wounds by peoples on the Baja peninsula.

In the late 1960s researchers screened elephant tree extracts for bioactive compounds in the hopes of identifying substances that could inform the development of new antitumor drugs.[11] They isolated a compound called deoxypodophyllotoxin, a natural product now known to possess activity against not only tumors but also inflammation and viruses. Unfortunately, the compound is toxic in animals, which has limited its usefulness in drug development. Some of its derivatives may yet hold promise, however. A second compound, beta-sitosterol, similar in chemical structure to cholesterol, was also isolated from the elephant tree. Studies conducted since have suggested that the compound may be active against diabetes and oxidative stress and may be able to reduce cholesterol levels. It has also been studied as a means of relieving symptoms of benign prostatic hyperplasia in men.

The status of the elephant tree is uncertain. In the Anza–Borrego Desert there may be just over two thousand individual trees. The California Native Plant Society lists the species as rare, threatened, or endangered. It is otherwise considered to be secure in the United States, despite the unknown status of many of its populations. Its relatively limited distribution in Arizona and its possible dependency on the gray vireo, which may be in decline, call its presumed stability into question. Furthermore, Arizona once was home to fragrant bursera (*Bursera fagaroides*), also known as torchwood copal, which actually represents two subspecies, *B. fagaroides* var. *elongata*, from Arizona, and *B. fagaroides* var. *purpusii*, from Mexico. In Arizona, it was found in the Baboquivari Mountains in Pima County. However, no extant populations of *B. fagaroides* var. *elongata* have been seen in decades. The last time a specimen was collected was in 1947. It is a popular species among collectors and can be cultivated, but its status in the wild is difficult to ignore when one thinks of the elephant tree. Could fragrant bursera's fate foretell that of the elephant tree?

There are numerous examples of coevolutionary affiliations shared by plants and animals, and many are of special relevance to the health of ecosystems and therefore to our own well-being. In many instances, in fact, if it were not for the magnificent and diverse interchanges that take place between plants and animals, there would be little for us to reap from nature, whether economically, recreationally, or spiritually. The production of seed crops, for example, depends on pollination by animals. Likewise, the biodiversity of grasslands is positively influenced by optimally sized herbivore populations.

The various coevolutionary relationships that exist between native species of plants and animals in North America provide us with an idea of the challenges facing rare plants on the continent. Unfortunately, many of those interactions and their significance remain only little understood.

Lost in the Wild

The Altamaha River originates at the confluence of the Oconee and Ocmulgee Rivers in central Georgia and winds gently southeastward, making its way down and across the state to the shores of the Atlantic Ocean. It is a relatively short river, just over 130 miles long, and its slow and graceful character seems fitting to the region, known for its peaches, its southern belles, and, historically, its steamboats, which plied the river's waters in the nineteenth and early twentieth centuries. Since then, the river has become a popular destination for those seeking recreation and leisure. But it has also become an endangered river, the seventh-most in the United States, a status attributed to a decline in water flow instigated by development, particularly the introduction of reservoirs and power plants.

Its endangerment is significant for many reasons, a major one being that the Altamaha River and its watershed, which covers nearly a quarter of the area of Georgia, support more than 120 different rare or endangered species. The river once hosted more rare species, including the region's only indigenous species of parrot, the Carolina parakeet (*Conuropsis carolinensis*). Arguably, its most famous and most mysterious resident was the Franklin tree (*Franklinia alatamaha*), a species that has not been seen along the river since the late eighteenth or early nineteenth century and that now bears the undesirable distinction "extinct in the wild."

The Franklin tree was discovered in Georgia in the fall of 1765 by celebrated American botanist John Bartram and his son William. The elder Bartram has been called the Father of American Botany. He was born in 1699, the third generation in a Pennsylvania Quaker family, and was raised southwest of Philadelphia, near Darby. He lost his mother in 1701, and his father passed away a decade later, leaving him to be raised by family acquaintances in the community. He later married and had two children but lost his wife and his oldest son. It was around that time, in 1727, when he decided that a change was in order. He purchased about 102 acres of land on the Schuylkill River in what was then Kingsessing Township, not very far from Darby. He remarried in 1729 and moved to the new property. There, he rose from the death that had so long surrounded him and with his new wife, Ann, eventually had nine children (one of whom they lost in childhood). It was also at the farm on the Schuylkill, and perhaps the same year that he remarried, that he saw new life bloom in the form of a garden on the sloping riverfront.

Bartram's garden occupied 6 to 8 acres of land, and he populated it with plants that he collected locally. He discovered a wide variety of species in the course of his collecting trips, and he gradually expanded his journeys, visiting ever more distant places. He was constantly on the lookout for curiosities, whether plant or animal, but it was his plant collection that garnered the most attention. In fact, the seeds of many of the plants that Bartram found eventually made their way to London, thanks to his friend and business associate, London cloth merchant Peter Collinson. Collinson was a British Quaker with connections to the community near Philadelphia. He also had a passion for horticulture. Bartram wrote to Collinson in 1733, initiating a relationship that grew to become immensely productive for the field of botany, even though the two never met in person. Bartram sent seeds and specimens regularly to Collinson, who in turn supplied materials to various collectors in Europe, some of them quite prominent, including Sir Hans Sloane, Carolus Linnaeus, and Philip Miller.

Bartram developed a significant customer base in Europe with Collinson's help. Collinson secured seed commissions for Bartram, and Bartram rarely failed to deliver. Furthermore, his expeditions became the botanical training grounds for his four sons, each of whom came to use their knowledge of plants in some way in their professional lives, whether as apothecaries, farmers, or collectors.

In advanced age, Bartram relied increasingly on his sons to run his business. After he passed away, his son John, Jr., who inherited the house, garden, and surrounding 140-acre farm, continued much of the business, sending large boxes of seeds from North American plants to Europe at high prices.[1] The younger John Bartram engaged in plant collecting to some extent but was kept at home for much of the time, overseeing the garden and dealing with the ramifications of war, which cut him off from European customers from about 1773 until 1783. After 1777 William conducted the business writing for the garden, often writing and signing letters for his brother, and he appears to have done much of the botanic naming and seed cataloging, likely because of his formal education and familiarity with Latin. William also spent many years traveling the southeastern United States in search of new finds. And it was he who was with his father that day when the history of the Franklin tree as it is now known began.

When the father-son team first spotted the tree, they were on their way to Fort Barrington, a station located on the banks of the Altamaha some thirty miles (by river) from the Atlantic coast. From there, they planned to head south along a well-traveled road that would take them to Florida, where they were to attend an Indian council.[2] After battling through swamps and stands of tupelo and cypress, a navigational miscalculation landed them a few miles downstream of their destination. Their misguided course, however, happened to take them through a quiet little grove on the northeastern side of the river, where they spotted the tree. The elder Bartram's diary entry on the events of that day contains the first documentation of the species, though he provided little detail, possibly due to the urgency of their travels.[3] His lack of specifics may have also been due to the time of year. The sighting was past the tree's flowering date, which left the Bartrams with little to glean in the way of the tree's taxonomic relationships.

The tree's curious appearance became embedded in William's memory. The next decade, he organized an expedition to the Altamaha in order to survey land that had belonged to the Creek Indians (the Muscogee). He eventually made his way back to the area where he and his father had spotted the unusual tree. He collected its seeds and saw it in full bloom. He would later write, in his influential work *Travels* (1791), that "it is a flowering tree, of the first order for beauty and fragrance of blooms." Its large flowers were "of snow white colour" with "gold coloured refulgent staminæ," and they "expanded themselves perfectly."[4]

It is thought that Bartram sent live specimens of the tree to John Fothergill, a British physician who had funded the expedition in return for specimens and seeds, illustrations, and field notes. That probably happened in early 1775, when several boxes of plants and shrubs appear to have been sent aboard a ship from Sunbury, Georgia, where William had spent part of the winter. The plants were forwarded to London, with their eventual destination likely being Fothergill's Upton Garden. Somewhere along the way, a specimen ended up at Kew Gardens in London. Several years later, Bartram gave two *Franklinia* seedlings to French diplomat Joseph Matthias Gérard de Rayneval when Gérard returned to Paris. *Franklinia* eventually made its way to the gardens at Trianon, and knowledge of the plant spread to European botanists, stirring the curiosity of collectors and growers.

In 1777, on the last leg of his journey through the Altamaha area and regions south, Bartram stopped off at the familiar grove where he and his father had twelve years earlier seen the tree for the first time. He collected a fresh batch of seeds and possibly some seedlings. Probably fleeing from the turmoil of war in the South, however, he left his major specimens, books, and probably some journals in Charleston, South Carolina, at the house of Thomas Lamboll. When he departed, it was likely that he carried only a small quantity of seeds, which he took back to the garden at his family's home in Philadelphia, now the site of Bartram's Garden, a national historic landmark. He returned in January 1777, before his father's death later that year. He likely started the seeds immediately upon his return, and John Bartram might have seen the first plants growing. And so their tree lived on. In 1781 botanist William Young, who was a neighbor of the Bartrams, observed flowers on 4-foot-tall *Franklinia* plants, probably the first time that the planted specimens flowered. Some claim that those plants sown in the Bartram garden are the ancestors of all Franklin trees under cultivation today.

For a long while, there was confusion about the tree's classification. William Bartram initially thought that the tree was a type of *Gordonia*, which are known for their white flowers with yellow stamens, similar to camellias in appearance. The tree later even acquired the species name *Gordonia pubescens*, among others. Renowned English botanist Sir Joseph Banks, who in the late 1780s had seen a drawing of the tree made by Bartram, was convinced that it belonged to *Gordonia*, though it is likely that he based his assessment on the name and description for *G. pubescens* that had been published by French naturalist Jean-Baptiste Lamarck.[5]

But it was not the name that Bartram came to assign the tree. After having compared its flowers and fruit with those of loblolly bay (*Gordonia lasianthus*), he realized that he was looking at two distinct types of plants. The flowers suggested that the tree, like *Gordonia*, belonged to the tea family (Theaceae). But it possessed other, very distinct traits. For instance, in contrast to *Gordonia*, it is a deciduous species. It also has shorter flower stems and wedge-shaped or angled wingless seeds, as opposed to *Gordonia*'s winged seeds. Its seeds further are positioned horizontally around the core of the fruit, whereas *Gordonia* has vertically positioned seeds, and the Franklin tree's seeds are greater in number relative to *Gordonia*. William Bartram named the new genus *Franklinia*, for Benjamin Franklin, who was a family friend. He used *alatamaha* as the species designation for the region where he and his father found the plant. (Alatamaha was an alternate spelling then in use.) Perhaps because of his American roots, John Bartram's cousin Humphry Marshall, also a botanist and Pennsylvania Quaker, used *Franklinia alatamaha* in his *Arbustrum Americanum* (1785), representing the first formal publication of the name. William Bartram used the same name later in *Travels*.

Interestingly, in the 1920s botanists in North America began to challenge the tree's placement in *Gordonia* in earnest. And in the following decades, there seems to have been consensus, at least among some, that the name given to it by Bartram was appropriate. So *Franklinia alatamaha* fell into wider use. What many had long suspected finally was confirmed by DNA sequence analyses published in the late twentieth and early twenty-first centuries, which revealed that the Franklin tree was in fact different from *Gordonia*. It actually was found to be somewhat closely related to *Schima* (Theaceae), a genus of evergreens native to tropical and subtropical parts of Asia. The flowers of the two are remarkably alike in appearance, as are their fruits and seeds, thus lending physical support to the genetic findings.

Following his return to Philadelphia in 1777, William Bartram never again saw the Franklin tree in the wild. Very few people who came after him saw it in its native habitat. One of the best accounts of the tree after Bartram was supplied by English nurseryman John Lyon, who saw the species in 1803. Lyon supposedly saw as many as eight plants. By that time, the species must have been in peril, with so few specimens to be seen.

The cause of the tree's extinction in the wild is a mystery. According to Bartram, the tree inhabited an area of 2 to 3 acres, which at that time would have required collecting in the extreme to decimate. So it is unlikely that collection had much of an impact. It is possible that the river, swollen from days of rain, may have flooded the tree out of its grove in 1796 during the Yazoo freshet. Or it may have been lost when land was cleared for settlement along the river. Alternatively, fire was common in the region in Bartram's time, and in a very dry year it presumably could have wiped out the Franklin tree. Disaster could have also come in the form of disease, possibly brought in with cotton, which began to be cultivated in the region around the time of the Franklin tree's disappearance.

Most likely, it was a combination of factors that contributed to the loss of the wild Franklin tree. It is also possible that by the time the Bartrams discovered it, the species was already close to extinction. Genetic analyses indicate that the species was a relict, which would mean that it had long been engaged in a struggle against the changes in climate that followed the retreat of the Laurentide Ice Sheet. Supporting that idea is the fact that species of the tea family have long inhabited the North American continent. Some 95 percent of tea family species that were established during the Eocene epoch (56–33.9 million years ago) in North America are now extinct. In addition, before the middle of the Miocene epoch (before 16 million years ago), the tea family tribes Gordonieae and Stewartieae may have mixed as they migrated over Beringia, the strip of land that once connected North America with Asia. The *Franklinia* line is thought to have diverged from the closely related tea family genus *Schima* by the Middle Miocene, suggesting that their split may have been effected as their tribe (Gordonieae) crossed the land bridge.[6] Similar to the Franklin tree, members of the genus *Stewartia* have limited distributions and could be considered on the verge of extinction.

With such an ancient existence on the North American continent, the fate of the Franklin tree leaves behind an eerie feeling of uncertainty for other relicts, including Fraser fir. And similar to Fraser fir, there is a sense of desperation, a sense that we need to know all that we can about the Franklin tree and its natural history. Perhaps that knowledge would lay to rest the great mysteries surrounding it, and maybe it would tell us something about the future of Fraser fir and other threatened trees in the southeastern United States.

We have only the scant details of diary entries and reports published in the nineteenth and twentieth centuries on the region where the Franklin tree existed. The portion of the Altamaha where the Bartrams observed the tree is characterized by sand hills and bogs, where the soil would have been acidic, with occasional overflows of the alkaline river water neutralizing the acidity.[7] The sand-hill bog habitat is also considered to be the type locality for the fevertree *Pinckneya bracteata*, a species that William Bartram recorded as being in close proximity of the Franklin tree. Interestingly, both species grow in well-drained acidic soils under cultivation. The Franklin tree, however, seems to be highly susceptible to root rot, easily succumbing to the disease where present in soil, and it can be fickle about soil pH levels and sun exposure.

Those characteristics seem to agree with the idea that the Franklin tree evolved on a coastal plain with a sandy barrens environment. Cultivated trees can withstand very cold winter temperatures and can thrive in the mid-Atlantic and into coastal New England. During the ice ages, the species may have been forced south into a southern refuge along the Altamaha River. At that time, when large amounts of water were locked in ice sheets and the coastal plain extended almost to the continental shelf, it is possible that the area in which the species could grow was much larger than the area that existed when Bartram discovered the tree. As the glaciers melted, however, the Georgia coastal plain was flooded. Flooding likely led to the confinement of location and eventual wild extinction of the Franklin tree. The species is highly susceptible to changes in soil pH, and flooding with neutral pH water can kill plants.

It was not long after William Bartram's time that the land along the Altamaha became significantly altered, primarily through land clearing. Still, extensive searches were conducted in the nineteenth and twentieth centuries in hopes that a population might still exist somewhere. One of the first expeditions of note was performed by botanist Henry William Ravenel, who from spring to fall of 1881 made no fewer than five trips to the region in search of the tree. Ravenel apparently went at the urging of Charles Sprague Sargent, who was then director of Harvard University's Arnold Arboretum. Prior to Ravenel's attempts, others had searched the area but to no avail, which was the same result for the various collectors and nurserymen who tried to hunt down the plant after him.

Edgar T. Wherry, who was a specialist in soils and plants, made at least four expeditions to the region, all without success. Naturalist Francis Harper and botanist Arthur N. Leeds explored the area around Fort Barrington in 1933, and three years later Harper returned, that time by way of the Old Post Road, a route that had been used by the Bartrams. Harper was an avid researcher of John and William Bartram's journeys and discoveries. For him, traveling that road probably was like stepping back in time, for though surely changed by modernization, to one so steeped in the Bartrams' work, its significance must have clung vividly to his imagination. A month after that initial drive, Harper had Leeds follow the route with him. From the road they were unable to catch much of a glimpse of the Franklin tree's sand-hill bog habitat. The next year, they returned again, bringing others with them, including four individuals of the Bartram family. Their area of exploration had been reduced by elevated water levels in the river, and like all who had come before, they were unable to find the tree.

William Bartram wrote that the Franklin tree obtained a height of 15 to 20 feet. In cultivation, it grows from 10 to 20 feet, sometimes higher, and may live for many decades. It often has multiple vertical stems or trunks, and it is covered with an interesting white-striped, ridged bark that grays with age. The 3-inch-wide flowers blossom in succession in July and August, though flowering time may be earlier or may extend later, depending on local climate. Its fruits are large and round and have a woody pericarp, inside of which are ten or so cells that contain the seeds.[8] While Bartram remarked on the beauty and fragrance of the Franklin tree's flowers in

Travels, he neglected to mention the brilliant color of its leaves in autumn, possibly because he never saw the tree along the Altamaha late enough in that season. But indeed, along with its late summer flowering, its fall color is among the traits most valued by horticulturists. The oblong, dark green, glossy leaves turn orange, purple, or red and provide great contrast for flowers that might linger into fall.

While it is difficult to know what lessons are to be learned from the Franklin tree's extinction in the wild, primarily because we cannot be sure of the cause of its loss, there is a message in the legacy of its discovery. John Bartram handed down his knowledge of plants to his sons, and he inspired them through his explorations and passion for plants. Their interest must have been heightened by the opportunity to join their father on some of his outings, which necessarily involved hiking, camping, and horseback riding. Their regular contact with plants, whether in the family garden or on collecting trips, inspired an everlasting appreciation for nature. That appears to have been true particularly in William's case. He, too, passed on his

Figure 6. The leaves and flower of the Franklin tree (*Franklinia alatamaha*). (Cr Kara Rogers)

knowledge and love of plants, keeping the family tradition alive in the education of Ann Bartram Carr, the daughter of John Bartram, Jr. In 1812 she inherited the family house and garden, and from that time, with her husband, Robert Carr, she tended the garden plants and maintained the seed trade.

Under their tenure, the Bartram family garden flourished, ultimately housing some fourteen hundred native species of plants in addition to a number of exotics. By 1847 financial problems began to jeopardize the Carr property, and three years later the couple was forced to sell. Railroad industrialist Andrew Eastwick became the new owner. The property was maintained under Eastwick, and he seems to have preserved it well. It was only after his death in 1879 that it began to fall into neglect. But even then, John Bartram's reputation lived on, and his family's devotion to the garden must have been infectious, because the city of Philadelphia approved its preservation. In his will, Eastwick left control of the property to the Pennsylvania Company for Insurance, which in 1893 and again in 1897 sold portions of the land, including the house and garden, to the city. Thereby was formed Bartram's Garden, which since 1893 has been managed primarily by the John Bartram Association.

Americans continue to benefit from that preservation effort, and the Bartram family's love of plants and nature lives on. It is something to think that one man's passion for exploration and botany has transcended centuries and inspired not only his children and descendants but also absolute strangers. That legacy is a testament to the value of knowing nature, a value that is not tangible in a material sense. Rather, it is a kind of value that adds depth to our lives. It gives us a connection to the past, where the secrets of our origins and the clues to who we are as a species, like those of the Franklin tree, lie buried in the passage of time. It also gives us something to pass on. Sharing our knowledge and our interest in nature with our youngest generations helps to ensure that an appreciation for the natural world persists. That is why the story of the Franklin tree matters.

To be lost in the wild implies the existence of cultivated individuals (or captive, in the case of animals). Franklin trees have been cultivated since William Bartram's time, so we know much about what the species looks like and what soil and moisture conditions favor its growth. To be entirely extinct, however, is quite another situation. In those cases, we often are left with only written words and, if we are lucky, illustrations. In the case of *Thismia americana*, we are left with two scientific papers, some slides, preserved specimens, and now a few photos of those specimens, but that is all.

There are parallels between the story of the Franklin tree and that of *Thismia americana*, but there are also vast differences. The two species hail from geographically and ecologically distinct regions, and they differ in their fundamental physical and biological characteristics. The Franklin tree, with its eye-catching white flowers, striated bark, and fall color, is alluring and graceful. *Thismia americana,* by contrast, was a strange subterranean life-form that forced its tiny flower through the earth above only once a year. Preserved specimens look reminiscent not of plants but of alien beings, with worm-like roots diverging from a tapered central body, the top portion of which looks not unlike a tadpole. Add to that its unusual life history, about which little is known, and the mystery that often haunts the extinction of plants deepens. Plants have a tendency to disappear with remarkable subtlety and for reasons often unknown. *Thismia americana,* which was discovered in August 1912 by Norma E. Pfeiffer and which vanished without a trace five years later, illustrates that subtlety and mystery in exceptional fashion.

Pfeiffer was a graduate student at the University of Chicago when she spotted *Thismia americana* in Solvay, a patch of prairie just south of Chicago near Lake Calumet. She was at the site to gather liverwort specimens for a class she was scheduled to teach that coming fall at the University of North Dakota. She was concerned that the school lacked the types of plants needed for the class, and so, in the company of a friend who would also be serving as an instructor that fall, Pfeiffer made several collecting trips to Solvay.

The process of taking up liverwort specimens requires one to be on one's hands and knees. Had Pfeiffer not been so low to the ground, she might have never noticed *Thismia*, a possibility that projects an air of improbability over her discovery. The plant's tiny, pale white flower strained to clear the soil surface, standing about half an inch off the ground and measuring about the same in diameter. The rest of the plant, including its leaves, was hidden underground, entirely out of sight. So the plant would have been very easy to miss, and in fact Pfeiffer's friend, who was looking around in the same area for liverwort, overlooked it.

Pfeiffer knew that she was looking at something not only beautiful but unusual too. The almost translucent white of the flower, with blue-green accents in the perianth (petals and sepals), was unlike anything else found in the moist prairie habitat that then characterized the Chicago lake plain region. The perianth traced the outline of a bishop's mitre, with three of the sepals and petals, of approximately the same length, arching upward and inward toward one another to form something like a basket handle over the floral axis. The other three were erect or bent gently backward. Pfeiffer did not know right away that she had found a new species of plant, much less a new genus in North America, but she carefully gathered samples to take back to the University of Chicago. It was soon after decided, following discussion between Pfeiffer and her professors, that she toss out her previous thesis problem and embark on a new one that focused on describing and studying the odd little plant.

For the next year, while in North Dakota, she examined in painstaking detail the specimens that she had found on the Chicago prairie. But she was unable to find much of use at North Dakota's library to aid her in the plant's identification. As a consequence, a thorough exploration of its taxonomic position was not possible until her return to Chicago the next summer. Her professors suspected that the plant belonged to Burmanniaceae, a family native primarily to wet tropical forests.[9] The stark geographic separation from its relatives was puzzling. Pfeiffer's plant matched many others in the family in its morphology, lack of chlorophyll, and mycotrophic lifestyle, drawing nutrients from fungi attached to its roots. But what was a species in a tropical family doing on a prairie where winter temperatures could stay below freezing for weeks at a time and summer drought and fire were regular threats?

Adding to the mystery was Pfeiffer's discovery that the plant was similar particularly to other *Thismia* species in Burmanniaceae. At the time, *Thismia* was known to contain just one other temperate species, *T. rodwayi*, which had been reported in 1890 by German-born botanist, geographer, and physician Ferdinand von Mueller.[10] This plant, which sometimes goes by the name fairy lantern, had been discovered in Hobart, Tasmania, by Leonard Rodway, who wrote to Mueller about his find. After examining a specimen, Mueller was surprised to learn that it was a type of *Thismia*. Only a few other *Thismia* were then known, and they were all from tropical habitats. Since Rodway's discovery, *T. rodwayi* plants have been seen in eastern and southeastern Australia and on New Zealand's North Island. The species, however, is exceedingly rare.

An inclination for a tepid climate was not the only characteristic shared by Pfeiffer's *Thismia* and *T. rodwayi*. In both, the sepals and petals that form the mitre-shaped perianth are either readily separated, without the need to be torn apart, or are fused along their ridges at the mitre apex, with the tips free and briefly interwoven. While there are variations on that theme—for instance, some *T. rodwayi* may have fused inner perianth lobes—in general, the "free" inner perianth arrangement is different from other *Thismia*, which tend to possess fully fused inner perianths. It is a technical detail, but the perianth structure of *Thismia americana* and *T. rodwayi* has been described as intermediate along the spectrum of possibilities in the genus.[11] Dutch botanist Frits P. Jonker, who prepared a monograph of the Burmanniaceae in 1938, thought that the two species might have been one and the same, though others since have suggested that *Thismia americana* may not be closely related to *T. rodwayi*. After all, the latter is a taller plant, reaching almost 3 inches in height, and its flowers differ in color, becoming salmon pink when fully exposed. Nonetheless, it remains the best candidate for *Thismia americana*'s closest relative, at least of known, living *Thismia* species.

When Pfeiffer realized that her plant was in fact a type of *Thismia* new to science, she effectively accelerated her graduate career. Her discovery enabled her to complete a doctoral degree in botany in 1913. At the age of twenty-three she was the University of Chicago's youngest doctoral graduate up to that time. Her doctoral research was summarized in "Morphology of *Thismia americana*," which appeared in the February 14, 1914, edition of the *Botanical Gazette*. The summary is one of few descriptions of the form and structure of the species, and no other source comes close in detail.

After completing her degree and joining the biology department at the University of North Dakota, Pfeiffer continued to study the plant. Every summer through 1917 she returned to Chicago to observe *Thismia americana* in its wild prairie home. She made her last trip to Solvay in the summer of 1917 but did not have enough time to conduct an extensive search. In the little time that she had, she was unable to find her familiar subject. It would have been impossible for her to know then, but the empty search meant that her last sighting, in 1916, would become the last time that anyone would see *Thismia americana* alive.

What Pfeiffer had learned about the plant's reproduction she communicated in a paper in 1918.[12] She suspected that it was pollinated by an insect, given the arrangement of its flowers. After its petals fell off, it left behind an inconspicuous, cup-shaped fruit that measured just over one-tenth of an inch in diameter. Its seeds were impossibly small, about one-hundredth of an inch long and less than half that in diameter. Many species in Burmanniaceae are known to self-pollinate, and in *Thismia americana*, pollen tubes possibly even penetrated the stigma. But cross-pollination has been described in Burmanniaceae, too, suggesting that some species rely on both strategies.[13]

Some of the specimens that Pfeiffer collected are kept under lock and key at the Field Museum in Chicago, while a few others are maintained elsewhere. They are the only specimens in existence, and it appears that

Pfeiffer made no further attempts to find the plant or to collect specimens following her trip in 1917. In the ensuing years at the University of North Dakota, she turned her attention to lilies. She later moved to the Boyce Thompson Institute for Plant Research in Yonkers (later Ithaca), New York, where she honed her expertise on lily biology and breeding.

Despite her transition in focus, however, Pfeiffer did not forget about *Thismia americana*. In 1948 she sent a map with directions to the location where she had spotted the species to Floyd Swink, an apprentice in the Field Museum's botany department. The directions took Swink and his mentor, Julian Steyermark, to a spot near 119th Street and Torrence Avenue. There they observed other plant species that Pfeiffer had described as associates of *Thismia*. Among those species in the low, moist prairie were black-eyed Susan (*Rudbeckia hirta*), calamus (*Acorus calamus*), common boneset (*Eupatorium perfoliatum*), giant goldenrod (*Solidago serotina*, or *S. gigantea*), narrow-leaved goldenrod (*S. tenuifolia*, later known as *Euthamia tenuifolia*), redtop (*Agrostis alba* var. *vulgata*, later known as *A. gigantea*), Shreve's iris (*Iris versicolor*, or *I. virginica* var. *shrevei*), swamp milkweed (*Asclepias incarnata*), and meadow spike-moss (*Selaginella apus*, or *S. apoda*), as well as the low-growing, nonvascular plants *Aneura pinguis* and *Hypnum*. But while its associates were there, *Thismia americana* was nowhere to be found.

Swink and Steyermark wondered whether a more extensive search would be fruitful, so in September 1951 botanists with the Field Museum conducted a search for *Thismia americana* in the location specified by Pfeiffer's map. They, too, however, were unable to locate any specimens. But though it turned up empty, the search for the plant was related in a January 1952 article in the *Chicago Sun-Times*.[14] According to Linda A. Masters, who in a 1995 paper in the journal *Erigenia* summarized the events relating to *Thismia americana* in the decades following Pfeiffer's discovery, someone who knew Pfeiffer sent her a clip of that story. Shortly thereafter, Pfeiffer wrote to Field Museum botanist Theodore Just, who had participated in the 1951 search, explaining the dates of her sighting of *Thismia americana*. Many thought that it had been observed only from 1912 to 1914, but Pfeiffer noted sightings for July 1915 and September 1, 1916. She also explained that she thought that the plant was still in the region, a hunch based on her suspicion that it could hibernate underground. Unfortunately, before another search could be organized, the site indicated on Pfeiffer's map was overtaken by fill.

More than three decades after Pfeiffer had written to Just, she wrote to Illinois botanist Robert H. Mohlenbrock. She mentioned that in 1914, when she returned to the original site where she had discovered *Thismia americana*, she found a barn in its place. It was in that letter that Pfeiffer mentioned that she had also seen the plant about a third of a mile distant, between prehistoric beach ridges of the ancient glacial Lake Chicago. At that location, she found it growing among cattails (*Typha*), suggesting a possible second habitat type for *Thismia americana*.

Pfeiffer's notion regarding underground hibernation offers a key means of explaining *Thismia americana*'s presence in North America. If the plant could hibernate for long periods, then it could survive in a temperate climate, despite the tropical origins of its genus. Over the course of thousands of years, the species could have diverged from its Asian ancestors and migrated over the land bridge, similar to the course taken by the Franklin tree. Every summer, or perhaps over longer intervals, the plant popped up through the ground and spread its seeds. And so by short bounds it could have eventually made its way onto the North American continent, finally settling in the moist prairies of the Midwest, perhaps finding the dampness there reminiscent of its ancestral habitat. That is only a hypothesis. In reality, it is not known how the species actually arrived here, much less why it would have settled in the prairie when it presumably could have migrated southeast, to humid forests or at least to places that rarely see freezing temperatures. *T. rodwayi* likely also diverged from its ancestors as it migrated to Australia, Tasmania, and New Zealand. And yet it ended up an inhabitant of primeval forests, where it grows in damp humus at the base of trees, presumably in a fashion similar to its tropical relatives. Furthermore, it is not known whether the region where Pfeiffer found *Thismia americana* actually was the species' preferred habitat. It is possible, in other words, that those habitat areas were outliers and that the species preferred the moist forests that once existed nearby.

In some ways, *Thismia americana*'s habitation of the moist prairie might not be as strange as it seems. The prairie that once characterized that region of Illinois rested on the exposed floor of the ancient glacial Lake Chicago, from which Lake Michigan was later born. Lake Chicago experienced repeated fluctuations in its levels, dropping and then rising, only to drop again. Those fluctuations contributed to a series of ancient shorelines that now are apparent in the form of sandy beach ridges, spits, and bars in the exposed lake bed. By eleven thousand years ago, with the retreat of the Wisconsin glacier, the last glacier to advance over the region, Lake Chicago was superseded by the slightly lower but much larger Lake Algonquin, which connected the basins of modern Lake Huron and Lake Michigan. Over time, the height above sea level gradually declined, and eventually, as drainage ways opened up, the Algonquin receded, giving way to Lake Chippewa and, by three thousand years ago, Lake Michigan.

The ancient lake plain is close to level, with Lake Michigan just feet below it. It is underlain primarily by sand and clay hardpan, which readily collects water. The plain is also crossed by the Chicago River. When the river flowed into Lake Michigan, it drained the lake plain, but sluggishly. As a result, low-lying hardpan areas of the region had (and still have) a tendency to flood, filling with water following extensive rainfall and snowmelt and giving those parts of the region a marsh-like character. The elevated sandy beach ridges (generally less than 10 feet in height) that lined the hardpan depressions, on the other hand, remained comparatively dry.

Marsh habitat interspersed with sandy, beach-like habitat presumably were features of the ancient plain, too, associated with periodic changes in the levels of Lake Chicago. The variability in habitat meant that the region likely was host to a diverse range of plants, with boreal and wetland species living only short distances from one another. Some of those species, including many herbaceous ones, came from the continent's Atlantic and Gulf coastal plains and found suitable living space in the various sandy, peaty, and gravelly habitats, giving rise to disjunct, or discontinuous, populations of their species. Examples of such populations found specifically in the northwestern Indiana Great Lakes region include eastern blue-eyed grass (*Sisyrinchium atlanticum*), slender yellow-eyed grass (*Xyris torta*), and little floating bladderwort (*Utricularia radiata*).[15]

The variety of plants that split off from their primary population centers to settle in the Great Lakes region suggests that such a move for *Thismia americana* would not have been extraordinary. The marsh-like qualities of the region could have been especially appealing for *Thismia*, assuming that soil moisture content was a determinant of its survival. But while we have a pretty good idea, for example, that eastern blue-eyed grass came to the Great Lakes region from the Atlantic coastal plain, we know virtually nothing about where *Thismia americana* came from. The current best guess is that it came from Asia. Whether it also took up residence someplace between there and Chicago is unknown. When it arrived is a complete mystery, too.

But Pfeiffer's suspicion about underground hibernation is enough to set the imagination soaring. The ability to survive beneath the ground until conditions are amenable for flowering would mean that the plant could very easily escape notice. Such an existence is not without precedent. The western underground orchid (*Rhizanthella gardneri*), for example, was discovered in the late 1920s in Australia only when a farmer noticed a sweet smell emanating from a small split in the ground in his garden. When the orchid's fragrant flowers bloom, they create minute cracks in the soil surface, apparently enabling their pollination by tiny insects. Maybe *Thismia* has a similar lifestyle. That would mean that new plants could emerge in new places without our ever knowing it. Likewise, remaining underground in a dormant state for years, with aboveground flowering for just one or two summers, would make the plant nearly impossible to find. *T. rodwayi* can live and flower beneath the soil surface, so such a habit would not be out of the question for *Thismia americana*. And if that really is how *Thismia americana* lived, then there is no knowing the true extent of its range or whether or not it is extant or extinct.

Because there are so many uncertainties about *Thismia americana*'s habits, some people are not ready to give the species up to extinction. Searches were conducted sporadically from the 1950s. From 1991 through 1994 professional and amateur botanists in the Chicago region gathered in the prairie remnants in Calumet and areas nearby for the annual "*Thismia* hunt." In 1991, when the first hunt was held, Chicago officials were heavily involved in planning another airport for the region. Among the locations that were under consideration was a site in the Calumet region, which happened to be in the same area where *Thismia* was first found. Although by that time the region was mostly industrial wasteland, sprinkled throughout it were natural areas that still supported significant populations of rare and specialized plant species. Amateur botanist George Johnson proposed the idea to look for *Thismia* in the Calumet area, doing so as a way to garner public awareness for the remaining natural areas. Johnson approached Linda Masters and Gerould Wilhelm, botanists then working at the Morton Arboretum near Lisle, Illinois. Masters and Wilhelm organized the first four hunts, making sure to invite decision makers from agencies such as the Illinois Department of Natural Resources, the US Army Corps of Engineers, and the US Fish and Wildlife Service. Many of the areas that were explored for *Thismia* have since become part of the region's natural preserve system, and plans to build an airport there have been abandoned, at least for the time being. A twenty-year anniversary *Thismia* hunt was held in 2011, despite the species' having been federally listed as extinct in 1995. In the course of searching for *Thismia*, volunteers discovered a number of species previously undocumented in the region, but they never found *Thismia*.[16]

Around the same time that the first *Thismia* hunts were taking place, other botanists affiliated with the Morton Arboretum set out to identify potential *Thismia* habitat sites in the Chicago region. In July and August 1993 they explored thirty-eight sites, five of which were thought to be of high potential and four of moderate potential in supporting the species. The five high-potential sites included the Sand Ridge Nature Center, Burnham Prairie, Calumet City Prairie, Dolton Prairie, and Wentworth Prairie. The sites were characterized variously by marsh, sedge meadow, wet–mesic prairie, or wet prairie vegetation, and they were home to some of the associate species described by Pfeiffer. Nevertheless, after two hundred hours of searching, the team found no evidence of *Thismia americana*.[17]

The 1993 survey brought to light some new ideas about possible causes of *Thismia americana*'s disappearance. Since Pfeiffer recorded little about the species' ecology, botanists since have had to draw on their knowledge of ecology in the region generally to figure out what might have happened. One possibility is that fluctuations in the level of Lake Michigan changed the plant's habitat. When levels were high, flooding in *Thismia*'s habitat could have been prevalent, whereas the retreat of lake levels could have caused its habitat to dry out. Maybe a rapid change from frequent wetness to frequent dryness was enough to decimate *Thismia americana*. It is likely, too, that the species was already on the way out by the time that Pfeiffer discovered it, with changes in natural fire regimes or some form of human disturbance disrupting its habitat. Human factors, particularly development of the prairie region, probably had a significant impact on the species' survival.

We might never know why *Thismia americana*, like the Franklin tree, disappeared. But strangely, we do have that sliver of hope. Maybe it is hiding underground someplace in the Chicago region. That in itself would be amazing, given the near-complete conversion of prairie habitat to urban jungle and suburban sprawl. If someone ever does find the species, it would be one of the most remarkable rediscoveries in the history of modern botany.

Mead's Milkweed

Historically, one of North America's largest ecosystems was its native prairie, which cut a wide swath through the middle portion of the continent. It stretched from Alberta, Saskatchewan, and Manitoba in the north to Texas and the northern edge of Mexico in the south, and in the west it unfurled in a graceful descent at the eastern edge of the Rocky Mountains, cascading all the way to western Indiana. That vast expanse encompassed a staggering 400 million acres, and it included the entirety of the Great Plains. Within it could be found three distinct prairie types: tallgrass, mixedgrass, and shortgrass.

Early European settlers saw the prairie in all its glory, where lithe grasses and colorful wildflowers surged and swelled in a gentle breeze and covered mile after mile of gently rolling land. The prairie seemed to go on forever, and to the early traveler there was a sort of comfort that came with the relative ease of traversing it. There were wooded draws and ravines where ephemeral streams provided moisture to support small stands of trees in which one could find shelter, and on the tallgrass prairie, where the grass grew above a man's head, man and horse could hide in plain sight. But grass that tall was seasonal, and trees in general were sparse. And the rolling green landscape could keep many travelers and residents on edge. It was open, and storms and Indians materialized with little warning. So while one had the advantage of seeing for great distances, there often was little physical protection to be found.

That changed following settlement, of course. People learned to cope with the elements. They built houses, and the Indians were forced from

75

the land. Seemingly rid of the landscape's dangers and brimming with the promise of prosperity, increasing numbers of settlers were drawn to the prairie and the plains. The land was swiftly converted to agriculture, and along the way, the prairie lost its ability to convey to us the sublime sense of endlessness that it did to those who first crossed it. Americans today live with that legacy, a heritage of nature's loss on the prairie.

While the prairie has not been transformed entirely, many areas of it are close to extinction. The shortgrass prairie has seen the least severe decline of the three prairie types. Decreases in shortgrass range from 20 percent in Wyoming to almost 86 percent in Saskatchewan. In the case of mixed-grass prairie, Texas still houses almost 70 percent, but only 0.1 percent remains in Manitoba. The tallgrass prairie has fared even worse. At its greatest extent, just over 17 percent remains in Kansas. Most other areas are down to 0.1 percent.[1] It is to that last and most decimated of the prairies that the story of Mead's milkweed (*Asclepias meadii*) takes us, and it is in the backdrop of migration onto the lands that once made up America's prairie where we begin.

Mead's milkweed was discovered in 1843 by American physician and botanist Samuel Barnum Mead. Mead found it in Hancock County, Illinois, which sits at the western edge of the state, with the county's west border tracking the Mississippi River and lying across the water from the southeastern corner of Iowa and the northeastern corner of Missouri. At the time, the county was not yet two decades old, but it was part of the Military Tract, a collection of lands set aside as a reward for soldiers who had fought voluntarily in the War of 1812. The county was settled quickly, and by the early 1840s it had become one of the most populous and influential in the state. Interestingly, much of the county's growth was due not to veterans and their families but rather to an influx of Mormons in the late 1830s. Most settled in Commerce, which religious leader Joseph Smith renamed Nauvoo, or "beautiful place." Many Mormons abandoned Nauvoo beginning in 1846, two years after Smith's death. They decided to migrate to Utah, following in the steps of their new leader, Brigham Young. For them, the path west was opened the year that Mead discovered his milkweed, which was the year of the Great Migration, when hundreds of Americans set out on the Oregon Trail.

The timing of Mead's discovery with the westward expansion was not purely happenstance. Mead himself may have been attracted to Hancock as a place of business, a county in need of medical doctors. In that era, medical doctors needed some knowledge of wild plants as sources of medicine, hence Mead's familiarity with botany. The promise of botanizing on what was then the frontier may have presented an added bonus for Mead. With growing numbers of people in the area, the county's botanical description was ever more likely. With that growth too, however, came the plant's destiny, as well as that of the prairie. Few, if any, people would have predicted the fate of the prairie then based on the sheer vastness of what lay ahead of them. The prairie's demise was unexpected, but because of it, the story of Mead's milkweed became, in many ways, the story of American settlement on the prairie.

It is thought that the original range of Mead's milkweed included the tallgrass prairie found in northwestern Indiana, Illinois, Missouri, eastern Kansas, Iowa, and southwestern Wisconsin. Mead actually saw it himself in Missouri, and others reported to him on its presence in Iowa. Still others verified it by collecting specimens in the other states. Its preferred habitat in those places appears to have been upland, or elevated, tallgrass prairie with a soil moisture content leaning a little more toward dry than wet, or what ecologists describe as dry-mesic. On those sites, Mead's milkweed enjoys full sun exposure, a benefit arising from the habitat's susceptibility to drought and fire.

Dry-mesic upland tallgrass prairie houses populations of plant species that are specially adapted to drought and fire conditions. One such adaptation is a deep root system, with some tallgrass prairie plants anchored by roots that extend as many as a dozen feet below the ground's surface. Rather than deep roots, some species, such as Mead's milkweed, possess shallow tubers, which help conserve water and nutrients in dry-mesic soils. Other adaptations seen among various species in the dry-mesic prairie habitat include those that allow plants to reproduce or grow quickly, helping them find or maintain their space in the prairie.

Fire and the frequency with which it occurs appear to be of particular importance for tallgrass prairie vegetation. Frequent fire drives succession and stabilizes late-successional vegetation. For Mead's milkweed, the historic regularity of fire in the tallgrass prairie resulted in unique growth responses, seen primarily in improved growth rates among Mead's milkweed seedlings and juveniles following fire.

Populations of Mead's milkweed are able to thrive in different types of dry-mesic prairie habitats. For example, in Missouri, the species has been found in chert prairie, limestone/dolomite prairie, sandstone prairie, and shale prairie. It has also been identified in glades and barrens, including the remnants of the sandstone barrens of southern Illinois. Each of the prairie habitats in which it is found differs by degrees in its collection of plant species, which is determined in large part by soil moisture and mineral composition and whether the soils are thin or deep.

Across the various habitats, Mead's milkweed maintains a characteristic appearance. Adult plants possess ovate leaves that are 2 to 3 inches long and about .5 to 2 inches wide, being positioned oppositely on an erect, hairless stem that stands between 8 and 16 inches in height. The leaves bear a herringbone pattern of venation, and both the stem and the leaves appear waxy. When broken, the stem oozes a milky latex-rich sap for which milkweeds generally are named. Young Mead's milkweed plants are shorter than adults and have relatively slender leaves, making them especially inconspicuous and difficult to identify in the tallgrass prairie.

Adult plants produce a lone, nodding cluster (umbrel) of between five and fifteen fragrant green-white flowers at the top of the stem. In southern areas, flowers bloom in late May, whereas in the northern extent of the plant's range, they typically appear in June. Several weeks after the cluster emerges, the flowers develop into one or more slender seedpods, which grow vertically, pointing skyward. The pods attain their full size, sometimes growing to as many as 4 inches in length, by late summer, and each one contains a puff of feathery seeds, which reach maturity in early fall. The seeds are released when the pods dry out and split open.

Mead's milkweed has an exceptionally slow rate of growth and is perennial, with a long life span. An individual can take anywhere from four to fifteen years to mature into a flowering plant. The upper end of that spectrum is comparable to the age when some long-lived trees become sexually mature, and, similar to those trees, Mead's milkweed, once in its mature state, can live for many years. Some individuals may live many decades in undisturbed environments. There are other slow-growing, long-lived milkweeds, but many species, including common milkweed (*Asclepias syriaca*) and butterfly milkweed (*A. tuberosa*), can flower within two to three years of germination.

Its curious physical and biological traits, its vertical seedpod ranking high among them, make Mead's milkweed a rather remarkable plant. But it is easy to miss in the tallgrass prairie. At full height, it is dwarfed by big bluestem (*Andropogon gerardii*) and Indiangrass (*Sorghastrum nutans*), which are about 3 feet tall at their shortest and often grow to at least twice that size. Next to vibrant palespike lobelia (*Lobelia spicata*), prairie blazing star (*Liatris pycnostachya*), prairie gentian (*Gentiana puberulenta*), and yellow coneflower (*Ratibida pinnata*), Mead's milkweed's pale flowers might go unseen by the casual observer. To some observers it may even seem like just another weed, an undesirable plant growing where it is not wanted, its pointy leaves, sticky sap, and strange-looking pods simply out of place. Even some of its less well known plant associates, such as button eryngo

(*Eryngium yuccifolium*) and Illinois bundleflower (*Desmanthus illinoensis*), look maybe a little less weed-like.

Mead's milkweed, of course, is not a weed on North America's tallgrass prairie. It also is not the only one of its kind found throughout its range. Eleven other species of milkweeds occupy space in various of its habitats.

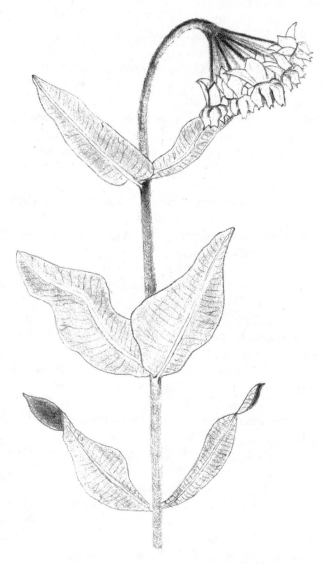

Figure 7. Mead's milkweed (*Asclepias meadii*). (Credit: Kara Rogers)

Those species include butterfly milkweed, common milkweed, narrow-leaf milkweed (*Asclepias stenophylla*), prairie milkweed (*A. sullivantii*), sand milkweed (*A. amplexicaulis*), short green milkweed (*A. viridiflora*), showy milkweed (*A. speciosa*), spider milkweed (*A. viridis*), swamp milkweed (*A. incarnata*), tall green milkweed (*A. hirtella*), and whorled milkweed (*A. verticillata*). Several of these species, such as butterfly milkweed, with its vivid orange-red flowers, and showy milkweed, with its oversized leaves and red-purple, star-shaped flowers, arguably are more attractive than Mead's milkweed. Short green milkweed often is confused with Mead's milkweed, though the former does not possess a single terminal umbrel. Several of these species also were prized by Native Americans for their medicinal value, whereas little is known about the particular medicinal or cultural significance of Mead's milkweed among the land's indigenous peoples.

Mead's milkweed seems to have garnered little notice from anyone except Mead and a few other botanists, at least initially. Mead formally mentioned the plant in 1846 in a catalog of Illinois plants published in *Prairie Farmer*, an Illinois farm magazine. He referred to the plant as "*Asclepias cordata* non Walt?" John Torrey, a leading botanist in America at the time, later received a specimen of Mead's plant and after examining it realized that he was looking at a previously unknown species. Northeastern Illinois University professor Robert F. Betz suggested in 1989 that Torrey assigned the plant a new Latin name, *Asclepias meadii*, sometime in the late 1840s. That name did not appear in published form, however, until 1856, in a work assembled by Torrey's protégé, Asa Gray.[2]

Within several decades of Mead's initial observation of the plant, the species had become rare. In December 1871 Mead wrote to Illinois naturalist Harry Norton Patterson, noting the situation and stating that the species had been "reduced very low" and that "perhaps the plough has destroyed it."[3] Mead's suspicions about the plow were warranted. By the time he wrote to Patterson, word had traveled about the rich soils in the "Prairie State," a nickname that Illinois had borne since at least 1842. But prior to the 1850s, farmers had struggled to plow the prairie. The massive root systems of its plants and the often dry and hard surface were at the heart of the problem. The cast-iron plows then in use were no match for the conditions. Blades bent and dulled after digging mere inches into the ground, and the animal strength needed to drag the equipment across the typically flat ground was amazing. In 1938 George A. T. Hise noted in an account on the history of progress in West Liberty, Iowa, that as many as six yoke of oxen—one dozen animals—were needed to slice a furrow about 1.5 to 2 feet wide on the prairie. An account of the history of plowing in Illinois by writer

James Krohe, Jr., that was published in 1981 mentioned the same.[4] Even with that much muscle power ahead of the plow, work was painstakingly slow, with only about an acre being turned up in half a day's work. Some farmers believed that soils that supported only grasses were inferior to those that supported trees, so for many, timber harvesting was an all-around more appealing affair than hitching up teams of oxen and scraping heavy sod off plow moldboards.

But by the 1850s, the prairie turned productive, owing to John Deere's invention of the steel-blade plow. Steel cut right through the dense root mats to expose the darkly colored silt-loam soil beneath, the soil that served as the primary substrate of the tallgrass prairie. John Deere lived in Illinois, and so Illinois farmers were some of the first to put the new plow to work. As its production increased and more and more of the steel-blade instruments made their way onto farmers' fields, the transformation of the Prairie State began in earnest. The same soon happened in surrounding Midwestern states.

Some called it "the breaking" of the prairie, as though the tallgrass was a wild horse to be tamed. Within decades, the essence of the prairie spirit, its native plant community, was shattered. The prairie's roots were torn, and its soil had been turned and turned. Vast expanses of tallgrass prairie were converted to farmland, and although initially a thriving enterprise, the perils of farming the prairie surfaced quickly. The first indication of difficulty was a reduction in crop productivity in the late 1800s due to the soil having been depleted of the nutrients and moisture that were essential to its fertility. Soil management programs were put in place, but that decline in crop production nonetheless continues to haunt prairie farmers to this day. Studies of virgin prairies, conducted primarily in Illinois, suggest that the topsoil layer on the tallgrass prairie was at one time more than 1.5 feet deep. But over the last 150 years, precious inches have been scraped and eroded away. In some places, more than half the depth of the topsoil layer has disappeared from converted tallgrass prairie.[5]

It is no wonder that many of North America's native prairie plants have suffered. The sensitivity of some of these species is emphasized by the rapid decline of Mead's milkweed, a species of naturally low abundance. Their continued peril is illustrated by the difficulty of restoring populations of Mead's milkweed in the wild. In 1879 Mead wrote that he had tried years before to cultivate the plant (presumably in his botanical garden) but had little success. At that time, he also made a plea for the plant's preservation. His words either were ignored or came too late, for a century later the species had been extirpated from several parts of its range. In 1988 the US Fish and Wildlife Service listed the species as threatened, and since then, a portion of the remaining populations has been monitored.

Historical information indicates that Mead's milkweed was found in forty-six counties in the six states that made up its original range. Natural populations are no longer found in two of those states, Indiana and Wisconsin, and in the early 1990s and again in the early 2000s, researchers reported that the species could be found in just one of the five counties in Iowa where it had been collected historically. In Illinois, it was extirpated from eight counties, including Hancock. While the discovery of previously unreported populations in Missouri and Kansas buffered those losses somewhat, by 2003 Mead's milkweed was down to thirty-four counties, spread across southern Illinois, south-central Iowa, eastern Kansas, and Missouri.[6]

The majority of populations, some 75 percent, are located on the Osage Plains in southeastern Kansas and west-central Missouri.[7] The Osage Plains represent a distinct physiographic region, in part because despite being free of glacier coverage during the last glacial period, the plains are gently rolling, a topography not unlike that of many glaciated areas. In both Kansas and Missouri, a few populations of Mead's milkweed also occur on glaciated sites, including in the Glaciated Physiographic Region of northeastern Kansas and the Glaciated Till Plain in northern Missouri.

In Missouri, roughly twenty populations of Mead's milkweed may be found in different areas of the unglaciated Ozarks. There the plant survives in places underlain by chert, limestone, and igneous rock. The topography of those areas also is notably diverse. In both the Ozarks and Osage Plains, most populations of Mead's milkweed occur in mowed meadows and pastures.

In Iowa, some populations of Mead's milkweed still exist on the Southern Iowa Drift Plain. The drift plain is known for its many glacial deposits that have been carved by stream erosion into steep, north–south-running ridges. In some areas, the terrain was so rugged as to turn the plow away, despite the exceptional quality of the soil for crop production. Still, only remnant prairies, providing limited habitat, exist there. On those prairies, the few remaining Mead's milkweed plants are supported by loess-covered loamy or silty-clay loam soils.

The disappearance of Mead's milkweed from most of Illinois is particularly tragic, since that is where modern botany's first encounter with the plant took place and where the tallgrass prairie first met the steel-blade plow. The last individuals of Mead's milkweed to inhabit the tallgrass prairie of Illinois were lost in 2001, when a plow destroyed the last population, which had persisted in a railroad right-of-way. Now just a handful of plants remains in Saline County in the Shawnee Hills. The Shawnee Hills is a rugged area in the southern part of the state that is noted for its rock formations, rivers, woodlands, and valleys. Mead's milkweed is found specifically in an open barrens on a sandstone escarpment within the Shawnee National Forest, where fire suppression may be hurting its chances of survival.

But while the numbers seem favorable—thirty-four counties, down just twelve from where our knowledge of the species began—of the populations of Mead's milkweed that are known, only a small fraction may actually be viable. In Illinois, Iowa, and eastern Missouri, for example, most plants are clones that were produced via asexual, or vegetative, reproduction.

Clonal individuals very rarely produce seeds and therefore generally are assumed to be nonreproducing. That unproductive state, which seems to be a last resort for population growth, comes about when Mead's milkweed is deprived of the opportunity for sexual reproduction. Such a circumstance arises, for example, when mowing cuts off flowering heads before they have fully developed. When this happens regularly, clonal populations become highly susceptible to a loss of genotypes, even though they may contain large numbers of stems. In general, sexual reproduction does not happen easily in Mead's milkweed. Like others of its kind, the species is self-incompatible, and thus outcrossing via pollination of genetically distinct individuals is required for the generation of viable seeds.

Scientists have been able to predict the future viability of Mead's milkweed in the various sites where it remains using measures of population size and habitat integrity. The predictions are broken down into rankings. Populations given an A ranking are those with high-quality habitat and in which sufficient numbers of young plants survive to maintain the existing population size. D-rank populations represent the opposite end of the spectrum, with habitat characterized by disturbance and the presence of nonnative and pioneer species and with population size or vigor markedly low. D-rank populations, as averaged over five years, consist of fewer than twenty-five ramets (individual clones) or fewer than one hundred ramets that do not produce viable seeds. According to the 2003 recovery plan, just six populations of the 171 total spread across the thirty-four counties could be given an A rank.[8] Five of those six occurred in Kansas. Of the remaining populations, the vast majority, some 103, were D rank.

Plants that reproduce sexually are able to reliably generate viable seeds with the help of pollinating insects. Insects that pollinate Mead's milkweed include digger bees (*Anthophora*) and bumblebees (*Bombus*). Betz speculated that digger bees might be the most important pollinators for Mead's milkweed, given the observed abundance of the insects in the vicinity of flowering plants. When the bees transfer pollen between plants in the process of outcrossing, they actively help maintain genetic diversity. That process is especially important for Mead's milkweed, which can so readily switch to vegetative reproduction.

When it comes to genetic diversity, Mead's milkweed is somewhat deceptive. Analysis of variation among allozymes, which are different enzymes produced by different forms (alleles) of the same gene, revealed a level of genetic variation on par with that of other species of milkweeds.[9] About 74 percent of genetic variation in Mead's milkweed was retained within populations, while variation between populations was about 26 percent, slightly better than other outcrossing species that are more widespread than Mead's milkweed.

At the same time, however, genetically distinct individuals, on which the success of sexual production depends, were found to be relatively few in number and tended to have greater distribution among populations than within populations. The lack of genetic diversity was especially pronounced for small populations, especially those derived from single clones, where a lack of sexual reproduction was suspected to have caused a loss of genetic possibilities. The loss of those possibilities was also significant among clonal populations in hay meadows, where annual summer mowing undermined sexual reproduction.[10] By contrast, burned sites appeared to house greater numbers of genetically different plants, as well as greater numbers of flowering ramets. The realization that plants at burned sites possess increased levels of genetic variation follows the observation that the growth of juvenile plants is improved by fire. Seedling survival is also elevated in burned areas, indicating that fire can have important effects on the demographics of Mead's milkweed.

Investigations of genetic diversity in Mead's milkweed have also implied that geographically distinct populations may not be genetically distinct. In other words, the alleles found in plants in Kansas may be the same as those found in plants in Iowa, Illinois, and Missouri. Genetic differentiation across geographic range has been found in other species of prairie plants, and so the possibility that such differentiation is not present in Mead's milkweed could be unexpected, especially given the species' occurrence at sites with different soil conditions. In general, plants in the southern extent of the species' range endure acidic soils that are sparse in nutrients, whereas those in the northern extent grow in nutrient-laden soils. So while Mead's milkweed has adapted to life under various conditions, that adaptability might not be reflected in a genetic differentiation of geographically isolated populations or in the plant's status in the wild. And neither is the species' status consistent with that of other North American milkweeds.

Why Mead's milkweed has suffered and other milkweeds have not remains a mystery, although several factors likely have contributed to its decline. Mead's milkweed tends to colonize successional habitats at a very slow rate, suggesting that the relatively quick large-scale disturbance to prairie habitats outpaced the species' ability to colonize suitable areas. Thus, its disappearance probably is linked in part to its apparent preference for relatively stable late-successional habitats, which are lost when prairies are disturbed. Moreover, that populations of Mead's milkweed are limited in number probably has very much to do with the fact that the species was rare to begin with. Small prairie remnants were unlikely to contain populations.

Factors affecting the insects that pollinate Mead's milkweed, including declines in pollinator populations and limitations in pollinator abilities, may have aggravated the plant's decline once that decline had been set in motion by habitat loss. The ability of bumblebees to retain milkweed pollen sacs for an average time of slightly longer than a day may help to explain how the plant's genes maintained their coverage over large geographic areas, at least prior to the fragmentation of its habitat.[11] With its populations now separated by relatively large distances, one wonders whether that same genetic coverage will persist.

Milkweeds, similar to many other flowers, pose certain challenges for pollinating insects and have a rather demanding flower anatomy. In fact, while the milkweed flower is wonderfully complex, the delivery of the pollen sacs in imprecise fashion causes pollination to fail. The roughly 150 different known species of milkweed, all classified in the genus *Asclepias* in the dogbane family (Apocynaceae), are distinguished from other species of plants particularly by their flowers.

In many milkweeds, the corolla (the petal collective) is reflexed, or bent downward, covering the sepals and exposing a central crown that is made up of five appendages, each of which essentially is an extension of a filament from one of five stamens. The five appendages are known as hoods, though they look more like little bowls, which is appropriate, since they serve up the plant's nectar. Projecting upward from each hood in some species is a horn. The curious crown structure surrounds the ovaries and the gynostegium, a fusion of the anthers and stigma. The gynostegium is a defining characteristic of the *Asclepias* genus and its close Apocynaceae relatives, and its structure plays a key role in insect pollination. When an insect lands on the crown, it tends to slide down into one of multiple slits that occur along the sides of the gynostegium. As the insect pulls itself up to escape, a thin cord running between pairs of pollen sacs, or pollinia, that are housed within adjacent anthers becomes hooked on its legs. When the insect flies away, it takes the pollinia with it, eventually depositing the sacs on the blossom of a different plant. *shelter — pollen dispersal*

In order for cross-pollination to be successful, the pollinia must be delivered to a plant of the same milkweed species, and they must arrive there in a specific orientation. That orientation is thought to develop as the cord connecting the pollinia dries during the insect's flight to the new plant. In that orientation, the pollinia must then land in an inconspicuous groove on a receptive stigma in the receiving plant. Only in the correct position can pollen tubes that develop from the pollinia penetrate the ovary.

Despite the challenge, the system is not inefficient in terms of resources or genetics. While seed production is limited by the need to cross among different genotypes and by resources and the presence of pollinators, seed-pod production is tightly regulated. It has been shown in other milkweeds, for example, that overpollination usually does not result in the production of additional pods, presumably because doing so would draw unnecessarily on valuable energy resources. Likewise, during drought years, seedpods typically are aborted.

With the combination of sexual and asexual reproductive strategies, milkweeds have been able to continue to exist despite threats to their habitats. But continuing to exist is different from thriving. It is hard to imagine what North America's prairies looked like when native prairie milkweeds and all the other prairie plants were at their peak. Historical paintings and photographs can provide some insight, but even by the nineteenth century the land had already begun to feel the settlers' presence. Those mediums also could capture neither the immensity of the prairie nor the diversity of its life. It is, for example, impossible to glean an idea of the sheer insect biomass that would have supported so many plants and flourished through feeding on them.

In 2003 the US Fish and Wildlife Service listed eight actions that would need to be carried out to recover Mead's milkweed.[12] Among them were protection and management of habitat, targeted increases in population size and number, field surveys to identify new populations or promising habitat where new populations could be introduced, research on techniques to help ensure the success of restoration, maintenance of populations currently conserved, public education, and progress review. The plan also described criteria that could be used to measure success. Recovery would be complete when twenty-one highly viable populations of Mead's milkweed were established across the species' historic range, including across the eleven physiographic regions of its range, with each population having enjoyed, for a period of fifteen years, stable or increasing growth. To be highly viable, a population had to consist of fifty or more mature, seed-producing individuals, as well as show signs of population growth and genetic diversity. High viability also required space, such that each population had to have an area available to it that covered at least 125 acres where the habitat had reached a late-successional stage, was managed with fire, and was granted long-term protection. If all goes according to plan, officials believe recovery will be complete by 2033.

The obstacles to reaching those targets, particularly population and habitat viability, are many, however. A major limiting factor in the recovery of Mead's milkweed is the minuscule amount of tallgrass prairie that remains in the plant's native range. Limited habitat means that only relatively small populations can be successfully restored. Too many plants restored to an area could result in outbreaks of disease or pests, ultimately undermining recovery efforts. The continued destruction of suitable habitat also poses a problem. Habitat identified as promising for the introduction of new Mead's milkweed populations could be destroyed before official restoration work begins. That has happened in the past, and there is little in the way to prevent it even now. Likewise, remaining populations on private lands could be lost at any time, and workers and volunteers cannot simply sow Mead's milkweed seeds and expect plants to grow. Studies have shown that juveniles transplanted to new sites from nurseries seem to perform the best when it comes to restoring the species. But nursery-raised plants might not survive transplantation into the wild, and if they do take, seeds will not be produced for another six or so years. Still, that is better than plants grown from seed, which have relatively low survival rates and mature at a very slow rate, taking two to three times as long as transplanted juveniles to achieve reproductive maturity.

The recovery of Mead's milkweed depends on an awareness and a thorough understanding of the challenges that lie ahead. Whether those challenges are surmountable remains to be seen. In 2012 the US Fish and Wildlife Service published a five-year review, which included an assessment of recovery criteria and whether efforts made up to that time had met those criteria.[13] A total of 330 populations of Mead's milkweed existed at the time of the report's publication, but very few of them were found to be highly viable. Those populations that did earn that ranking were concentrated within just three physiographic regions, and thus they did not meet the distribution requirements of the recovery criteria. Furthermore, viability scores of the various populations were described as preliminary, since information on variables such as habitat size and management and plant reproduction were lacking for most sites, thereby preventing complete assessment of all the viability variables at stake. The preliminary viability scores that were calculated suggested that highly viable populations were rare. Just three populations earned that designation. They were located in three regions: the Glaciated Physiographic Region of Kansas, the Osage Plains Physiographic Region in Kansas and Missouri, and the Ozark-Springfield Plateau in Missouri. Whether or not any of those three populations had experienced a stable or increasing trend for the fifteen years prior to review was unclear, as population monitoring had been inconsistent, yielding insufficient data for accurate assessment. The 2012 review further suggested that since some Mead's milkweed plants can take upward of twenty-five to thirty years to reach maturity, based on estimates from plants introduced at new sites, fifteen years might not be long enough to capture population trends. Extending observation out accordingly would surely delay the original projected date of recovery, pushing it back from 2033 to possibly 2050 or later.

Since the early 1990s, Mead's milkweed has been introduced at multiple sites. The first introductions took place in Illinois and Indiana, including at the Morton Arboretum's Schulenberg Prairie in northeastern Illinois and at the Biesecker Prairie Nature Preserve in northwestern Indiana. The introductions involved the planting of seedlings and one-year-old juveniles. Those early experiments revealed the improved viability of juveniles over seedlings in the wild and demonstrated the benefits of fire on seedling and juvenile growth. Some juveniles flowered after just two years, but other plants, particularly those that were introduced as seedlings, appeared to slip into a state of dormancy, with no indication of when they might achieve reproductive maturity. Similarly, by 2012 Mead's milkweed plants that were introduced to Wisconsin's Till Plains about a decade earlier had not yet flowered. None of the introduced plants, according to the 2012 report, had recruited young plants into their populations. With some of the problems known and characterized, researchers were hoping for a better outcome regarding introductions slated for the tallgrass prairie at Marais des Cygnes National Wildlife Refuge in Kansas.

Successfully reintroducing plant species is an immensely difficult process, even when problems unique to a species have been identified. Few reintroductions of long-lived rare or threatened plant species have succeeded, and still fewer have succeeded and resulted in improvements in the species' threatened status.[14] With that in mind, it seems optimistic to think that Mead's milkweed could be recovered within three decades, in part through introductions into suitable habitat.

But the combination of more time and steady population monitoring could pay dividends in the identification of highly viable populations. There also is the possibility that new populations of Mead's milkweed could be discovered. In 2009 researchers reported that Mead's milkweed had been identified on eighty-eight new sites in Anderson and Linn Counties in Kansas. The researchers also confirmed the existence of the species at six sites previously recorded. Two populations in Anderson County consisted of stems numbering in the thousands. All the plants inhabited areas where natural plant communities remained largely intact and where some extent of land conversion or habitat fragmentation by agriculture or other forms of development had occurred.[15] With those threats in place, the future of the newly discovered populations was uncertain, but the plants were there nonetheless. Other populations have been found in both Iowa and Missouri.

Perhaps the most compelling argument in allotting resources to the protection and recovery of Mead's milkweed is that the species is a prairie plant. The North American prairie is iconic. It is a part of America's identity. Every settler who ventured west of the Mississippi River faced the endless expanse of plains, and those settlers crossed the prairie. Their survival of that journey became a part of their character, the same way that the survival of Mead's milkweed in the face of habitat fragmentation by the plow is now a part of its character. Many more-robust prairie plants have persisted with seemingly little consequence, but there are others like Mead's milkweed whose status in the wild is tenuous. Among those plants are eastern prairie fringed orchid (*Platanthera leucophaea*), western prairie fringed orchid (*P. praeclara*), prairie bush clover (*Lespedeza leptostachya*), and leafy prairie-clover (*Dalea foliosa*). The first three of these species are listed as threatened in the United States. The last, leafy prairie-clover, is endangered. Each of those plants is inherently valuable, and in the story of each, we find a story in the history of the westward expansion of America, when European settlers discovered some of North America's greatest natural treasures. The prairie was one of those treasures, but it was quickly overshadowed by places such as Yellowstone, the Grand Canyon, and Yosemite. Compared with those natural features, the prairie was more tangible for humankind, something that people could not only inhabit but also farm and prosper from.

But we have now wrung from the prairie nearly all it has to give, and if we continue to place impossible pressure on its resources, it will fail. To prevent that from happening, we need to give back to the prairie some of what we have taken from it, and not just for the sake of the prairie. The Great Plains is North America's agricultural backbone. Americans and people beyond the borders of Canada and the United States depend on the food grown across the prairies of the Great Plains region. As a result, the state of our food supply is intimately connected to the well-being of prairie ecosystems.

Florida Torreya

The versatility of Mead's milkweed, in terms of its geographical distribution and ability to grow in prairie habitat of varying soil types, contrasts starkly with the narrowly defined range and uncertain habitat requirements of Florida torreya (*Torreya taxifolia*), a rare conifer confined to the eastern banks of the Apalachicola River in Florida's panhandle. In 1984 Florida torreya was listed as endangered under the Endangered Species Act, and it remains one of North America's most endangered species of conifer. The IUCN considers it to be critically endangered, one step removed from the designation of extinct in the wild.

Only within a forty-mile stretch of the Apalachicola and its tributaries between the southern edge of Georgia and Liberty County, Florida, does one have the opportunity to see the rare tree in its wild state. The odds of finding one unaided are fairly poor, as there are only an estimated five hundred to six hundred individuals left. Although the species' abundance prior to human settlement is unknown, the current population is more than 98.5 percent smaller than the population that was in existence before the 1950s, representing a rapid decline. The haste of its retraction is alarming, as is knowing that its loss continues and that its prospects for recovery in the absence of human intervention are exceedingly unfavorable. Florida torreya is a primarily dioecious species: each tree bears either male or female reproductive organs. But of those several hundred trees remaining in the wild, just six appear to be able to produce male or female cones.[1]

The limited natural range of Florida torreya is an example of what ecologists describe as endemism, when the distribution of a species or other taxonomic group of plants or animals is restricted to a given place. Fraser fir also is an example of an endemic species. Although endemic species may be abundant in their localized habitat, they are highly susceptible to endangerment from habitat destruction, disease, pests, and climate change.

Certain parts of North America are more inclined toward endemism and high levels of plant diversity than others, and the southeastern United States is one of those areas. There are at least six centers of endemism in the Southeast, and on that list is the Apalachicola region of the Florida panhandle.[2] There are several reasons that might explain why certain species of plants are confined to that region. For Florida torreya, those reasons have to do mainly with current and past climatic and human factors, which have not only shaped its habitat but also influenced its migration across the landscape.

Habitat in the Apalachicola region is characterized primarily by slope forests, pine flatwoods, and relatively treeless dry prairies, areas that are home to dozens of endemic plant species, many of which are rare. Florida torreya specifically is found within slope forests in the Apalachicola drainage, where the land is etched by bluffs and ravines, some of which are deep, canyon-like cuts known as steepheads.[3] The highest elevations in the habitat, the tops of the ravines, sit about 250 feet above sea level. The lowest areas are about one-fifth that height. Surveys in the 1980s indicated that most Florida torreya—almost two-thirds of the trees—existed in the middle elevations, between about 100 and 180 feet. Just under a third of trees favored areas above those elevations, while the remaining fraction occurred below them.[4] The growth of Florida torreya at mid- to low elevations suggests that it might benefit from moisture that seeps down from the uplands.

The climate of the region is one of combined cool and warm temperate conditions. From 1981 to 2010, the average minimum temperature in the Apalachicola region was about 58.6°F, and the average maximum temperature was 78.1°F. Annual precipitation over that period averaged 57.7 inches.[5] The amount of rainfall in the area contributes to the maintenance of a mesic habitat, one with moderate amounts of moisture. Moisture levels influence soil conditions, as do variations in topographic relief. Soils along the sides of deep ravines in the Apalachicola, for example, consist primarily of sandy red clay, whereas upland soils tend to be enriched with aluminum and iron or may have a thick upper layer of humus. Those characteristics mean that many different types of trees and shrubs can exist within slope forests, and for that reason, some consider the slope forests of the Apalachicola to be one of only a handful of biodiversity hot spots within the United States.

The region's biodiversity is reflected in Florida torreya's natural habitat, where the tree associates with a wide array of broad-leaved species and a few other conifers. Some of its most notable associates include beech (*Fagus grandifolia*), magnolia (*Magnolia grandiflora*), and white oak (*Quercus alba*), which in the summer months filter out much of the sunlight, leaving only diffuse rays to reach Florida torreya in the understory. Other trees in its habitat, several of which may also contribute to the canopy cover, include American holly (*Ilex opaca*), American hophornbeam (*Ostrya virginiana*), hickory (*Carya* spp.), loblolly pine (*Pinus taeda*), southern sugar maple (*Acer barbatum*), spruce pine (*P. glabra*), sweetgum (*Liquidambar styraciflua*), and tuliptree (*Liriodendron tulipifera*). Florida torreya's habitat is also home to various types of woody, climbing vines and shrubs. Common examples are crossvine (*Bignonia capreolata*), Florida yew (*Taxus floridana*), poison ivy (*Toxicodendron radicans*), and yaupon (*I. vomitoria*). Growing beneath the vines and shrubs are different kinds of sedges and grasses, such as giant cane (*Arundinaria gigantea*) and panic grass (*Panicum* spp.). Other low-growing plants include ferns and herbs such as American climbing fern (*Lygodium palmatum*), little sweet Betsy (*Trillium cuneatum*), and partridgeberry (*Mitchella repens*).[6]

In such a biologically rich environment, it is easy to think that Florida torreya and other endemic species of plants are thriving and have a secure future. And maybe some of those species will have a chance at persisting for some time yet, since habitat areas in the Apalachicola region have been modified rather than completely destroyed. Still, in addition to Florida torreya, various threatened and endangered species live there. Florida skullcap (*Scutellaria floridana*), Miccosukee gooseberry (*Ribes echinellum*), telephus spurge (*Euphorbia telephioides*), violet butterwort (*Pinguicula ionantha*), and white birds-in-a-nest (*Macbridea alba*) are listed as threatened in the federal record and as endangered in Florida. Meanwhile, Apalachicola false rosemary (*Conradina glabra*), Chapman's rhododendron (*Rhododendron chapmanii*), Gentian pinkroot (*Spigelia gentianoides*), and Harper's beauty (*Harperocallis flava*) are listed federally as endangered species. The outlook for most of these species, like that of Florida torreya, is poor.

There are some aspects of Florida torreya that are unexpected—for example, it is buried in an understory where broad-leaved trees tower over it, with limited conifer associates. The number of coniferous species in North America that are found in nonmountainous areas at such southerly latitudes is relatively small compared with the diversity of conifers seen growing in more northern latitudes or at higher elevations. That in itself adds interest to Florida torreya, and it raises questions. For instance, how did the species come to exist along the Apalachicola River? There are questions, too, about how the species came to its imperiled state so suddenly and within such a biologically diverse area. Though they may seem unrelated, these issues—how the species arrived in its present-day habitat and how it became so few in number—both factor into the effort to save Florida torreya from extinction.

There are six living species of *Torreya*. They are, in addition to Florida torreya, the California nutmeg tree (*T. californica*), Chinese nutmeg tree (*T. grandis*), Farges nutmeg tree (*T. fargesii*), Jack's nutmeg tree (*T. jackii*), and Japanese nutmeg tree (or kaya, *T. nucifera*).[7] Similar to Florida torreya, several of those species have seemingly unusual geographic occurrences, most notably California nutmeg, the only other North American species, which is limited to the Coast Ranges and the foothills of the Cascades–Sierra Nevada, with no populations between those two locations. The rest of the *Torreya* genus is found in parts of eastern, central, and southwestern China, with the exception of Japanese nutmeg, which is native to the mountains of Korea and central and southern Japan.[8]

The geographic distribution of the *Torreya* genus typically is described as disjunct. It is, in other words, discontinuous, with some species in eastern Asia and two others in distinct regions of North America. Rather than having a distribution that obviously agrees with existing theories of plant migration or the movement of major landmasses in the geologic past, geographically disjunct groups are more enigmatic. How some members of the group came to exist in a geographic pocket in one place, while others are found halfway around the world, with no other members between, can form quite a puzzle. That is true particularly when there is little fossil evidence available to fill those gaps, as is the case with Florida torreya and the rest of the *Torreya* genus. As a result, precisely how Florida torreya ended up in the state's panhandle is not known. But there are plausible theories, based partly on the limited repository of fossil evidence and partly on analyses of genetic relationships between the different *Torreya* species.

The *Torreya* genus appears to have a very deep history, going back farther than most other woody plants.[9] Fossils of leaf and shoot fragments that appear to belong to the genus were recovered at sites in Europe and were dated to about 170 million years ago, which places them within the Jurassic period, when many of the major groups of plants and animals that we know today began to emerge. The oldest *Torreya* fossils uncovered in North America were dated to roughly 100 million years ago, placing them in the

Early Cretaceous. More recent fossils, from the Tertiary period (about 65.5–2.6 million years ago), have been found in both Europe and western North America, suggesting that in that span of geologic time, *Torreya* may have enjoyed a fairly continuous distribution across parts of North America, Europe, and Asia.

Molecular evidence has revealed that the clade to which Florida torreya and California nutmeg belong diverged from the clade containing their Old World counterparts roughly thirty million years ago.[10] That timing corresponds with the widening of the Atlantic Ocean and the consequent separation of the North American and European landmasses during the Eocene epoch (56–33.9 million years ago), suggesting that populations of *Torreya* trees that were divided by the division of continents literally went their separate ways. In the few million years following the split, populations on each continent probably had begun to evolve distinct traits. And in both North America and Europe, *Torreya* appears to have maintained continuous distribution, meaning that there probably were few gaps, or at least not very distant ones, between populations. Following the divergence of New World and Old World populations, the latter in Europe began to migrate into Asia, a move likely made possible by the drying up of the Turgai Sea. Some populations eventually made their way to Japan. Around two million years ago, it is probable that *Torreya* in Europe began to go extinct.[11] The genus eventually disappeared from that region, leaving behind only those populations in North America and eastern Asia.

From about 2.6 million to 11,700 years ago in North America, populations of *Torreya*, similar to those of other conifers, were subjected to the repeated advance and retreat of glaciers. Perhaps at the beginning of that era, *Torreya* was widely distributed in the north, occupying the Circumboreal Region, which included the boreal area of North America and Eurasia and which was characterized by abundant forests. But as glaciation ensued, the climatic changes forced *Torreya* (and other boreal species) to migrate south. In Europe, the movement of the genus to more southern latitudes may have caused or expedited its extinction. In eastern Asia, the genus presumably moved into relatively low-lying areas. In North America, a southerly migration from the boreal region could have landed the genus in the areas of modern-day California and the Florida panhandle and possibly points between. With climatic warming and the northward retreat of the Laurentide Ice Sheet between 15,000 and 10,000 years ago, many boreal species began marching north again. Populations of *Torreya* in California and Florida, however, remained in their more southerly refuges. In California, the trees had at least someplace to go to partially escape warming climatic conditions—they migrated up in elevation, eventually adapting to the specific conditions of the Cascades–Sierra Nevada and Coast Ranges. Some have suggested that Florida torreya's adaptation to the Apalachicola region, influenced by the region's unique climate, soil, or topography, may have precluded its retreat north. Such a circumstance would not have been wholly unique to Florida torreya. Florida yew (*Taxus floridana*), for example, also took refuge in the region during the last glaciation and appears to have adapted to the region's climatic conditions. Alternatively, both species simply may have become trapped in the Apalachicola, trying to migrate away but failing and never fully adapting to the warmer conditions.

For Florida torreya, two other factors could have been at play, and either or both of them together could have prevented its migration out of the Apalachicola region. One is the elimination of large herbivores. Large plant-eating animals may have helped shape the landscape inhabited by Florida torreya, possibly influencing fire regimes, or may even have helped to disperse the tree's seeds. A second factor may have been human activity, which also would have altered natural fire regimes and may have led to changes in mammal populations that were vital to Florida torreya's successful migration. In the centuries that followed, then, it is possible that climate change and human activity worked in concert to steadily whittle down remaining populations.

Across the immense span of time in which *Torreya* has existed, scientific knowledge of the genus, and specifically of Florida torreya, has been but momentary, a brief flash at the end of a chronology that has lasted millions of years. Florida torreya came to the attention of American botanists sometime around 1833, when H. B. Croom noticed it growing along the eastern bank of the Apalachicola River in the region of Aspalaga Landing. Croom had leased a plantation near Aspalaga, which then was an important riverboat landing and which subsequently became a strategic post for the US Army during the Second Seminole War (1835–42). Locals knew the tree as stinking cedar or savine, names that reflected its pungent odor, produced when the foliage was crushed or the bark wounded or burned. But to Croom and the scientific study of plants, it was new.

Croom realized that the species had not been described or classified after evaluating certain features of the tree and comparing them with the traits of known species. Its undocumented nature meant that he had an opportunity to name the species. With apparently little hesitation, he decided that the tree should bear the name of one of America's most influential nineteenth-century botanists, John Torrey, hence, the genus name, *Torreya*.

Croom died shortly after his discovery, but before his passing, he had sent a branch, bare of reproductive structures, to English botanist Thomas Nuttall, who was familiar with North American plants, having spent much of his life botanizing in the United States. Nuttall considered the tree to be a type of yew from Mexico and classified it accordingly, giving it the Latin name *Taxus montana*. Torrey had also received several specimens, and his analysis suggested that the species shared certain characteristics with other plants of the yew family (Taxaceae) but was sufficiently distinct to warrant placement in a new genus. Scottish botanist George A. W. Arnott, who apparently had obtained plant material from Torrey, also evaluated the species' characteristics. He came to the same conclusion as Torrey. Arnott published his assessment in 1838 as "On the Genus *Torreya*" in the *Annals of Natural History*. With that publication, Croom's tree was formally introduced to science.[12]

That Florida torreya bore Torrey's name happened to be a detail that struck a nerve of curiosity among American botanists. Following Croom's discovery, many professional and amateur botanists in America had come to learn of the species. Some were familiar with the display of torreya sprigs by Torrey Botanical Club members. Others were intrigued by the existence of young transplanted trees, such as one in Central Park in New York, which was believed to have been started from a seedling sent by Croom to Torrey, who then forwarded it to a botanist in the city. But although knowledge of its existence had spread, few had ever actually seen the tree in the wild.

Asa Gray had also become curious about the new species. In the second half of the nineteenth century, he embarked on what he described as "a pious pilgrimage to the secluded native haunts of that rarest of trees, the *Torreya taxifolia*." His account of his journey, published as a letter to the editor in the *American Agriculturist* in 1875, is one of few known documents that transports readers into the tree's secluded habitat as it existed in that historical era. Gray's journey proved trying, but he eventually was rewarded with a sighting of Florida torreya. He knew much about the species before having made the pilgrimage, since he had been a student of Torrey when Torrey had received specimens from Croom. So Gray knew to also look for a small herb called croomia (*Croomia pauciflora*) growing beneath Florida torreya. Croomia's association with the tree in the Apalachicola region was a remarkable thing. A similar association had been described between *Torreya* and a type of croomia in eastern Asia. In the southeastern United States, the seeming uniqueness of the relationship led Gray to speculate that Florida torreya had once thrived in Alabama and Georgia, places where croomia occurred but where Florida torreya did not. Gray's speculation on the tree's historical distribution in the land led others to wonder about where the tree really belonged. The question of where its range truly lies is at the center of modern debate on how best to protect the species from extinction.

Gray also provided one other important glimpse into Florida torreya's future. In the areas he searched, he observed just one tree with forming "fruit."[13] And by the time of his visit to the Apalachicola, little more than three and a half decades after Croom first saw the tree, the celebrated stands of torreya that had once stood triumphant over the crest of Aspalaga Bluff had been cut down. Gray believed that the trees had been sacrificed for use as fuel for the steamboats that plied the river's waters. In his essay, he noted, too, that specimens that Croom had planted on his property in Quincy, Florida, had been destroyed and that the tree was used for fence posts. Botanist and physician A. W. Chapman noted the same usage and wrote in 1885 that the wood also had been used for roof shingles and other "exposed constructions." Both men understood that the tree was rare, vulnerable, and valuable, which combined to make its existence exceptionally precarious.

In a deep ravine off the bluffs at Aspalaga, Gray noted one specimen in particular, a torreya that he estimated to be about 4 feet around, the largest individual he had seen. The largest specimen on record, measuring almost 3 feet in diameter at breast height and having a sprawling, 40-foot canopy, was a tree planted in North Carolina. Most healthy Florida torreya in cultivation are much smaller, averaging about 1 to 1.5 feet in diameter. Some cultivated specimens occasionally reach a vertical height of around 40 feet. Historically in the wild, the species grew to about one and a half times that size. Most wild Florida torreya in existence today, by contrast, are but young, small trees, an inch or less in diameter and a couple feet in height, owing to devastation by a fungal blight.

When full grown (to about 40 feet), Florida torreya's numerous whorled branches create a magnificent conical canopy with a spread of about 20 feet. Each branch is adorned with lustrous green leaves, which are stiff, flat, and needle-like (sharp to the touch) and about 1 to 1.5 inches in length. Running parallel to the midvein of each leaf are two gray or reddish sunken bands. The leaves of Florida torreya are further characterized by their unusual arrangement. In conifers, leaf pairs may be decussate, such that successive pairs of opposite leaves crisscross one another, or they may be bijugate, such that successive pairs form a more narrow angle (i.e., less than 90°) to each other. Bijugate leaf arrangements are exceedingly rare in conifers but are found in both *Torreya* and *Cephalotaxus*. But whereas the bijugate pattern is constant in *Cephalotaxus*, Florida torreya and other torreyas have a biphasic arrangement. Leaf pairs are decussate in the production of bud scales but later become bijugate, in which successive pairs are positioned at an angle of about 68°.[14]

The bark of mature Florida torreya is relatively thin, being only about half an inch thick in mature individuals, and is carved by shallow, irregular grooves. In weathered individuals, it is dark brown to gray or black in color, often with hints of orange, and beneath it rests a yellowish inner bark. The outer bark of young trees, the only individuals to be seen in the wild, typically is yellowish brown.

Wild Florida torreya trees do not produce cones until they are about twenty years of age or around 10 feet in height. Male cones are small and characteristically crowd in around the axils of neighboring leaves. They are oval to oblong in shape, with four pollen sacs borne per scale. Male cones appear in March or April on the previous year's shoots. Female trees produce cones in the same months, but with ovules developing on shoots of the year. The ovules are contained in fleshy, sessile sacs that are solitary, scattered throughout the tree, and fewer in number than male cones. They are fertilized about four to five months after pollination by wind-disseminated pollen from male cones. In the second year, the ovule develops into an oval-shaped, roughly inch-long seed that is yellow to brown in color and that has a thick, woody seed coat with a meaty inner layer of white albumen. When mature, it measures about 1 to 1.5 inches in length and slightly less in width. The seeds mature in late summer and fall. The female cones of Florida torreya, along with the characteristics of its foliage, are used in the identification of the species.

Figure 8. Top-down view of a Florida torreya (*Torreya taxifolia*) seedling, showing leaf arrangement (*left*). Seed cones of Florida torreya (*right*). (Credit: Kara Rogers)

But while the general anatomical features of Florida torreya have been fairly well characterized, comparatively little is known about its life history, particularly concerning seed dissemination, germination, and seedling development.[15] The large size of the Florida torreya seed suggests that animals likely played an important role in seed dissemination in the past. Gray squirrels, for example, are known to eat the tree's seeds and to gather its seeds, making squirrels potential agents of seed dispersal. At the Biltmore Gardens in Asheville, North Carolina, squirrels gather the tree's seeds and bury them in middens. Seeds that are not recovered may eventually germinate. In the Apalachicola region, past human activities, such as hunting squirrels for food, may have reduced squirrel populations and consequently the animal's role in seed dissemination. Some have postulated that a now-extinct species of tortoise may have been an important agent of dispersal for Florida torreya.[16]

Once dispersed, Florida torreya seeds require afterripening, or essentially a period of rest before germination begins. During that time, the tiny embryo contained within the seed of Florida torreya matures. Afterripening helps to prevent seeds from germinating in unfavorable conditions, such as during a drought. At the time of germination, moisture may be critical for Florida torreya, based on the relatively high moisture levels in its environment. The site of germination is beneath the ground surface, where conditions tend to be humid in the tree's habitat. Germination occurs within one to three years following the completion of embryo development. The total time required for a fertilized embryo to become a seedling can span as many as five years, taking into account the time needed for afterripening.

The growth habits of Florida torreya seedlings in the wild are largely a mystery. Seedling establishment and early growth may be aided by mycorrhizae, or associations between fungi and plant roots. Seedlings may also benefit from some amount of shelter and shade and may require moist, well-drained soils to grow. In shady areas, they may outcompete other species for space, although too much shade may be detrimental. Their initial growth rate, when healthy, may be typical for intermediate-sized conifers. So, between eight and twelve years of age, they may be anywhere from 6 to 8 feet in height. They continue to grow following sexual maturation.

Florida torreya also uses vegetative reproduction to regenerate, particularly when a tree has suffered damage to its aboveground parts. Sprouts can shoot up from the roots, from the lower part of the main trunk, or from the root neck, which lies at soil level or slightly above or below. Although multiple sprouts may grow on parent trees, only one will survive, and then usually not for long, as most succumb to fungal blight.

The dieback of Florida torreya is thought to have begun just before the start of World War II. One of the first to report an observable loss of the trees was Florida state extension forester L. T. Nieland. In 1938 Nieland "observed a general unthrifty condition of *Torreya* which worsened with defoliation."[17] The vulnerability of the species was noted at about the same time or shortly thereafter by Florida botanist Herman Kurz, though Kurz saw no evidence that the trees were dying. Even a decade and a half later, when Kurz journeyed with two botanical parties into Florida torreya habitat, he saw nothing unusual about the trees. For a time, then, it appeared that Kurz was correct and that Nieland's observation was an outlier. The only threat to Florida torreya, it seemed, was deforestation by humankind.

But in 1955, officials at Torreya State Park, in the heart of Florida torreya habitat, reported a steep decline in the tree's populations. In 1962 Kurz and fellow botanist Robert K. Godfrey surveyed three sites, one in Jackson County, one at Aspalaga, and one at Rock Bluff. At all three, they were able to find "only a few abortive sprouts," as they recounted in a letter published that year in the journal *Science*.[18] It turned out that what Nieland had observed was real, but seemingly overnight the situation had turned dire. Adult populations had been decimated, and there was no indication that the species was reproducing. Florida torreya, Kurz and Godfrey warned, was on the brink of extinction.

The dieback was attributed to a fungal disease that had been observed on needles and stems. The mature needles of affected trees displayed small, circular lesions, which were tan in color and which eventually expanded or coalesced, causing needles to become necrotic. The disease then spread into the stems, resulting in the formation of cankers. Needles, having been killed, dropped to the ground. Entire twigs were left bare. Some wore on their tips only small bunches of young needles, which were initially immune to the disease but once mature would succumb as well. Necrotic stems and needles became home to fungal fruiting structures, which perpetuated infection. The disease seemed to be worse for trees that received full sunlight than for those in more shaded areas. But the result was still the same. Trees were severely defoliated. And without cones, adult Florida torreya were left reproductively dead.

The only individuals that were spared the disease were seedlings younger than six months. They presumably were started from seeds left behind by dead adults. By the 1990s, however, virtually all adult trees in the wild had been killed, and between 1,000 and 1,350 juveniles remained. The largest of the young trees stood less than 6.5 feet tall and had from one to five sexually immature stems that arose from a single root axis. Many young trees displayed symptoms of disease, and over time their stems were killed off, one by one. The likelihood of their survival being prolonged depended in part on the size of their main stem. A larger main stem meant a more promising outlook, at least for a while. But once the main stem died or when a tree was down to three or fewer stems, each a foot and a half tall or less, death ensued.

By 2010, among wild populations of Florida torreya, only six plants were able to produce cones. Two of those individuals were female trees at Torreya State Park. One was a female tree on property belonging to the Army Corps of Engineers at Lake Seminole, Georgia, and three were male trees on privately owned land. Much of the species' population consisted of stump sprouts, individuals that had reproduced vegetatively.

The identity of the fungal pathogen believed to be behind the decline of Florida torreya was unknown. Where it came from also was unclear. In 1967 S. A. Alfieri, Jr., and colleagues isolated multiple fungal organisms from affected plant parts.[19] Among their isolates were *Rhizoctonia solani* and *Sclerotium rolfsii*, as well as species of *Macrophoma* and *Sphaeropsis*. None of the organisms, however, produced symptoms of the disease when inoculated into six- to twelve-month-old seedlings. The researchers also observed a fungus known as *Physalospora*, which occurred with *Macrophoma*, but they were unable to establish pure cultures of it and so could not test it in inoculation experiments. Nonetheless, the unusual association of the two fungi led the researchers to suspect that *Macrophoma* and *Physalospora* were perhaps together responsible for stem and needle blight in Florida torreya. Nothing definitive came from that suspicion.

In the 1980s Alfieri and colleagues described other fungi from infected leaves and stems.[20] They included fungi in the genera *Alternaria*, *Botryosphaeria*, *Diplodia*, *Fusarium*, *Phyllosticta*, *Phytophthora*, and *Pythium*. Experiments by another group of researchers suggested that *Fusarium lateritium*, which was found to cause necrotic leaf spots in Florida torreya, was the culprit. But *F. lateritium* failed to produce stem cankers in experimental studies.

Some believed that the disease was caused by the introduction of a non-native pathogen, such as *Phytophthora cinnamomi*, a water mold known to cause root rot. Still others suggested that the frequency of leaf spot and stem cankers on Florida torreya, combined with the number of different organisms that had been isolated from infected trees, pointed toward some underlying cause, such as stress brought on by drought, which would have increased the trees' susceptibility to any number of infectious organisms.

However, the species had apparently survived previous periods of stress without succumbing to fungal disease.

The hunt for a specific causative fungal organism continued. In the 1990s the cankers of infected trees were found to contain a fungus known as *Pestalotiopsis microspora*. The organism, which was known for its ability to digest polyurethane, happened to be a natural resident of the inner bark of Florida torreya. The thinking was that *P. microspora* was opportunistic, becoming pathogenic when trees became stressed. But inoculation experiments indicated that the organism was incapable of causing stem mortality. Another lead involved a fungus in the genus *Scytalidum*, but it, too, turned out to be an unlikely cause. Still other researchers explored some of the soil-borne pathogens first reported by Alfieri, including *Pythium, Phytophthora, R. solani,* and *S. rolfsii.*

Interest in *Fusarium* also resurfaced. In 2009 a previously undescribed *Fusarium* species was isolated from Florida torreya and subsequently was found to produce canker symptoms in cultivated trees, ultimately killing them. The organism was later named *Fusarium torreyae,* and it seemed possible that it was responsible for at least some proportion of mortality in wild Florida torreya.[21] Some researchers think that it is in fact the outright cause of stem and needle blight in Florida torreya.

It is possible that the survival of Florida torreya is affected by interactions between various species of fungi and oomycetes that have been found in the roots and in the soil surrounding the roots of Florida torreya. Researchers have isolated organisms belonging to more than two dozen genera of fungi and oomycetes from the rhizosphere and root tissues of declining trees. Possibly also affecting the survival of Florida torreya is damage to protective communities of mycorrhizal fungi. Within the tree's native range and in trees transferred to forest or garden sites, researchers have identified a variety of arbuscular mycorrhizal fungi that associate with Florida torreya.[22] Nearly all of its mychorrhizal associates belong to the genus *Glomus,* which contains a number of species that help defend trees against root pathogens. The abundance of those fungi, however, appears to be greater among garden explants than among trees in native habitats. Wild Florida torreya currently inhabit heavily shaded areas, which, combined with the presence of disease, may limit the trees' ability to support beneficial mycorrhizal fungi.

But while the speed at which Florida torreya's decline occurred and the prevalence of diseased plants indicate that a pathogen (or several) is behind the dieback of Florida torreya, there has been speculation that other factors likely contributed to the species' loss once its decline had been initiated.

Perhaps the most significant of those proposed factors were logging and the conversion of upland pine forests into managed pine plantations, which may have led to alterations in the hydrology and structure of slope forest habitat. Nearly half of all small Florida torreya trees have also been affected by damage from deer browsing, and in some areas more than half may be affected by deer rub. In severe instances, which emerge mainly during the rutting season, deer rub damages the cambium and results in stem and tree death.

The ecological and cultural significance of Florida torreya lends a special importance to the species and to the effort to save it. The tree's designation as one of the world's most endangered conifers is something that many people would like to erase. So, despite the lack of specifics on the precise cause of Florida torreya's dieback, various organizations and botanical gardens, academic researchers, and citizen volunteers have become deeply involved in trying to give the species a future.

Much of the restoration effort has centered on a recovery plan that was developed by the US Fish and Wildlife Service in 1986 and reviewed in 2010. The plan listed seven actions that would be key to the species' recovery. Among them were protecting the species' existing habitat, controlling its decline, producing seedlings and cuttings, learning more about its ecology and life history, establishing experimental collections outside its native range, storing seeds, and reestablishing the species within its native habitat.

Wild populations of Florida torreya reside both inside and outside preserved lands. The three main areas of preservation are the Nature Conservancy's Apalachicola Ravines and Bluffs Preserve and Torreya State Park, both of which are in Florida, and the lands owned by the Army Corps of Engineers in Georgia. Efforts to protect existing habitat at Torreya State Park have included the prevention of erosion in slope forests, the restoration of uplands adjacent to ravines, the use of enclosures to protect trees from deer, and the control of nonnative species. Some private landowners planned to implement similar management strategies and proposed to establish experimental plots to test different recovery treatments. Examples of treatments that warranted further investigation included the use of fungicides or lime to control fungal disease, which could ultimately help control Florida torreya's decline. Of particular interest for fungicide treatments were agents that might be effective against *Fusarium torreyae*. If the fungus was the cause of the disease, treatments targeted against it might be able to effect a cure. However, because fungicides pose risks to mycorrhizal associations, on which Florida torreya may depend, and can be hazardous to animals and other plants, the use of fungicide treatments required extreme caution.

The process of producing seedlings and cuttings of Florida torreya centered largely on identifying seed sources, disseminating seeds for storage and propagation, and establishing seedling production programs. Although exceedingly few trees in the wild produce seeds, seed production is relatively good among cultivated populations. Seed-bearing trees have been identified at the Atlanta Botanical Garden and Callaway Gardens in Georgia, as well as at North Carolina's Biltmore Gardens. In high seed-production years, some five hundred to six hundred seeds may be gathered at the Atlanta garden, which houses the largest collection of Florida torreya outside its natural habitat. The Biltmore Gardens has also seen years where seeds number in the hundreds. While some of those seeds are preserved in seed collections, others are used for propagation or are given to other institutions or groups for use.

Seed-storage programs have met with limited success, however, and seedling and cutting production programs remain under development. The process of establishing disease-free seedlings and cuttings is not straightforward. Rooted cuttings from wild trees may still carry disease, and those from wild or cultivated trees may experience plageotropic growth, in which lateral growth dominates, resulting in a shrub rather than a tree. To circumvent the propagation of diseased seedlings, cuttings from nursery specimens may be used. For wild specimens, researchers have experimented with somatic embryogenesis, in which embryos are removed from fully developed seeds and then cultured to encourage the genesis of multiple embryos. The latter step takes advantage of the natural process whereby plants produce multiple embryos (only one of which normally survives). Somatic embryogenesis ultimately results in the generation of multiple copies of whole, disease-free plants.[23] Meanwhile, plageotropic growth in rooted cuttings may be overcome through the use of material cut from tree tops rather than from lateral branches.

Reproductive offspring of Florida torreya that have been produced at the Atlanta Botanical Garden are among the best candidates for the reestablishment of Florida torreya in the wild. The offspring were produced from the seeds of parent trees that were established as part of the garden's collections in the late 1980s. The parent trees were themselves established from cuttings originally collected at Torreya State Park by botanists with the Arnold Arboretum. Some of the seedlings from the Atlanta garden were reintroduced in 2002 into ravines from which the species had disappeared. Just over one-third of them were alive the following year.

Arguably, the most interesting aspect of the recovery plan for Florida torreya concerns the establishment of experimental collections beyond the boundaries of its habitat in the Apalachicola region. Specific locations have been chosen for the species' establishment in controlled settings and in wild Appalachian forest communities in both Georgia and North Carolina. The approach was originally premised on the observation that Florida torreya trees that had been planted at the Biltmore Gardens and other garden locations in the Appalachian Mountains appeared to be healthy and thriving.

The purposeful movement of Florida torreya into the wild in the Appalachians is an example of assisted migration, in which humans help species to become established in new habitats, typically to enable them to escape environmental threats such as climate change. Some researchers have argued that the Florida panhandle was the peak-glacial range of Florida torreya, as opposed to its actual native range, which they believe encompasses the southern Appalachians, the Cumberland Plateau, and possibly even areas farther north and west.[24] The argument for its survival in the Appalachians is based in part on the discovery of macrofossils in North Carolina and Georgia that were identified as *Torreya* and dated to the Late Cretaceous. It is also based on the general northward migration pattern of conifers following the last glacial period. Thus, the perspective that Florida torreya belongs in more northern habitats is based on "deep time," in which we dig back into the species' history in North America to determine its range and assist in its conservation.

In recent years, assisted migration of Florida torreya in North Carolina has been led by some members of the Torreya Guardians, a group of botanists, naturalists, and other individuals who share an interest in the protection of biodiversity. While some Torreya Guardians are academic researchers, others are private landowners or citizen naturalists. All participants are volunteers.

Not all Torreya Guardians support assisted migration, but to those who do, the recovery of Florida torreya is a process of "rewilding." From that perspective, trees that are cultivated in gardens are in effect caged, whereas those planted in forest habitat in the Appalachians or points farther north are simply repopulating their natural home areas. The difference is this. In Georgia, experimental populations of Florida torreya have been established at Smithgall Woods State Park and at Vogel State Park, both of which are located in the northern part of the state. Many individuals planted at Smithgall Woods have thrived thus far. Some have even reached reproductive maturity, producing annual crops of male and female cones. It is wonderful progress, but the trees have been grown in a grove and have been tended more or less as garden plants. They are not, in other words, surviving entirely on their own, nor were they scattered about the habitat as they might have been through a process that mimicked squirrel or tortoise seed dissemination. Rewilding of Florida torreya is further focused on private lands rather than public lands, owing to the nature of the Torreya Guardians as an unaffiliated group and to the idea that private lands host nonpristine settings, places where the land has already been disturbed, thereby circumventing possible ethical issues concerning the introduction of the species to lands where it would not otherwise occur in this millennium.

The first rewilding of Florida torreya into forested habitat on private lands that was led by the Torreya Guardians took place in 2008 near Waynesville, North Carolina. Volunteers planted thirty-one seedlings into two sites. The plants had been obtained from a North Carolina nursery and were grown from disease-free seed. Five years into the experiment, some trees were thriving, whereas others had died or were in a diminishing state. Several had died from damage caused by voles, which had chewed into the bark and cambium layer. Other factors, such as amount of sunlight and wet, heavy snowfall, may have influenced the survival or death of other specimens. Those volunteers involved in the assisted migration effort were analyzing the successes and failures in North Carolina to see whether they yielded information on Florida torreya's preferred habitat. Over the same period of time, a number of other Florida torreya seedlings were planted on private lands in North Carolina, Georgia, Ohio, and even Wisconsin.

The success of some of the North Carolina seedlings reaffirms what many people had believed following the success of Florida torreya specimens at Biltmore Gardens. Florida torreya was first brought to Biltmore in the late nineteenth century, according to correspondence between Chauncey Beadle, Biltmore botanist and horticulturist, and Charles Sprague Sargent, director of the Arnold Arboretum in Boston. In 1939 Beadle collected other Florida torreya trees from the Apalachicola and planted them at the gardens. He chose to plant them alongside a stream in a small ravine at an elevation of about 2,200 feet, where the trees have since gracefully endured subzero temperatures. It is not coincidence that some Florida torreya specimens have survived in North Carolina's mountains. The other species of *Torreya* survive in mountainous areas, having migrated up to higher, cooler ground when confronted with climate warming at the end of the last glacial period. Seeing as there were no mountains for Florida torreya to immediately climb up, its only option to escape warming temperatures would have been to move north and then up, into the southern Appalachians. To survive, researchers have projected, Florida torreya would have had to track the cool climate along the Chatahootchee River into the southern Appalachians, completing a journey of more than 370 miles.

It never made that trip on its own, of course. It never traveled north, and thus it was never forced to adapt along the way. But we can make the journey for it, delivering it to the Appalachians to help it escape extinction. The question is whether or not we should. The intentional movement of species by humankind is a highly controversial issue in conservation science. Species that are moved to areas where climatic or habitat conditions are deemed to be more suitable, for example, could become invasive in their new homes if they are particularly successful there. Florida torreya has a very low likelihood of becoming invasive in the Appalachians or in other habitats outside the Apalachicola region, and indeed, many introduced species have but small ecological impacts. But those that do become invasive can have significant effects on ecology and economy.[25]

The fate of species that are targeted for assisted migration and the fate of the ecosystems into which the species are to be introduced are other uncertainties. Florida torreya is close to extinction in the Apalachicola, and it could very well remain on the verge of demise despite assisted migration. There also is a risk of pathogen introduction associated with the movement of Florida torreya cuttings from areas affected by disease to areas that are unaffected. That risk has necessitated the use of seedlings grown from seeds produced by parent trees that exist outside the species' current range in the Apalachicola region. As a result, the seedlings are at least one generation removed from their wild ancestors. In addition, some seedlings from parent trees grown in nurseries in the Appalachians are better suited than others for life in the mountains. It is possible that those individuals differ genetically in slight but important ways from the wild Florida torreya of the Apalachicola region.

The argument for Florida torreya's movement north is also based on assumptions about its ancient range and preferred habitat. But no one can say one way or the other whether Florida torreya belongs in the southern Appalachian Mountains. If the assumptions are incorrect, much time and money will have been spent in trying to move a species to a place where it will not survive.

Perhaps science needs to go through a trial of assisted migration with Florida torreya. It is, after all, the only way to find out whether the process truly works, and it would help us to better understand advantages and draw-backs. As those involved in the Torreya Guardians assisted migration pro-ject have noted, the places where Florida torreya is being repatriated have already been altered by human activity. They are in a state of recovery, and so is Florida torreya. Given its low invasive potential, the risks of moving ahead with its assisted migration are minimal. The Guardians have also used a very cautious approach, and one grounded in science, if premised on assumptions. The data they have collected on Florida torreya habitat preferences, germination, and seedling growth are themselves of remark-able value. That information could fill a book—something not remotely possible before the project began.

Still, the thought of humans moving species around is unsettling, and those who are opposed to assisted migration and particularly those who are engaged in it must sense that. For, even though it is carried out with the best intentions for nature, assisted migration is yet another means by which humans can manipulate the environment. And so we may find ourselves returning to the question: How much human manipulation of the environ-ment is too much? Florida torreya is not the only species to exist in a refu-gium that has become inhospitable. There are many others, and they may someday need our help to migrate to more suitable habitats. Our repeated intervention could dramatically alter landscapes, especially in a region such as the Appalachians, which could receive a variety of endangered migrants from more southern areas. But if we do not intervene, endangered species trapped in their peak-glacial range could end up like the Franklin tree—extinct in the wild, the few remaining specimens "caged" in botanical gardens or scattered about on a few private properties. The Franklin tree existed in what is now described as the Altamaha refugium. And while the precise reasons for its extinction are unclear, its disappearance likely was aggravated by the retraction of its peak-glacial range, a tiny pocket of habi-tat along the Altamaha River.

In some ways, assisted migration is an opportunity to right past wrong-doings to nature of humankind. In the process, however, we cannot turn our backs on the places that affected species currently call home. So it is imperative that federal agencies, state parks, and nongovernmental organi-zations continue their vital work toward the recovery of Florida torreya in the Apalachicola. The species can be considered for downgrading to threat-ened status once five populations, each with sexually reproducing indi-viduals, have been established in protected parts of its range. It could be delisted entirely when fifteen self-sustaining populations are established in protected ravine systems. All of those populations could emerge full and healthy in the Apalachicola, if a means of overcoming stem and needle blight is found. Habitat protection for populations in Torreya State Park and other preserved areas would be guaranteed. And so Florida torreya would live on, and it would do so in a place that it maybe should have departed from millennia ago.

Fickeisen Plains Cactus and Acuña Cactus

The Colorado Plateau is a high-elevation tableland that covers southern and eastern Utah, northern Arizona, northwestern New Mexico, and western Colorado, encompassing an area of 130,000 square miles. Incised by steep-walled canyons and eroded by the Colorado River and its tributaries, it is a deeply carved and weathered land, the effects of time borne openly on its surface. It is topographically a maze of basins, buttes, canyons, cliffs, escarpments, and mesas, and it is geologically a layer cake of red sedimentary rock. Much of the plateau, owing to its high elevation and geographical position, is dry, extremely warm in the summer, and very cold in the winter.

The climatic conditions on the Colorado Plateau make for a harsh existence for vegetation, which is relatively sparse across much of its area, with the exception of verdant valleys. But across the plateau, the climatic, geologic, and geographical characteristics have helped give shape to unique assemblies of plant life. As a result, the plateau houses numerous endemic species, one of which is the endangered Fickeisen plains cactus (*Pediocactus peeblesianus* var. *fickeiseniae*).[1]

Fickeisen plains cactus was listed federally as an endangered species in 2013, its populations on hilltops and along canyon rims and ridges in northern Arizona having declined. Also added to the country's endangered list that year was acuña cactus (*Echinomastus erectocentrus* var. *acunensis*), a species limited to southern Arizona's western Pima, Maricopa, and Pinal Counties, all of which lie south of the Colorado Plateau. While not the only species of cactus whose future is in jeopardy (more than two dozen others are recognized federally in the United States as threatened or endangered), as recent additions for protection under the Endangered Species Act and as members of Arizona's unique flora, they are of special interest. Few people have heard of either Fickeisen plains cactus or acuña cactus, and fewer still are aware of either species' plight. Indeed, like many rare, threatened, or endangered cacti, Fickeisen plains cactus and acuña cactus have escaped public notice. For more than three decades, they also managed to escape the notice of those who grant federal protection to plant and animal species, despite repeated reviews that identified the need for the species' protection.

The story of Fickeisen plains cactus begins in the mid-1950s, when Florence Fickeisen and Mr. and Mrs. Denis Cowper independently collected the species. Fickeisen found it northwest of the Grand Canyon, whereas the Cowpers collected it from a site that lay to the east and a little farther south, by the Little Colorado River. Having learned of the Cowpers' discovery, cactus specialist Lyman D. Benson proceeded to gather specimens from an area near Cameron, Arizona, to the east of the South Rim of the Grand Canyon and close to the Cowpers' collection site. Benson examined the plant closely and identified its defining traits, which he described in 1969 in the first formal account of the species.[2]

Over the course of the following decades, more came to be known about the plant's distribution and its population size. Its populations were found to be limited to northern Arizona, primarily to parts of Coconino County and its neighbor to the west, Mohave County. In 1977 the cactus was found at Sunshine Ridge on the Arizona Strip, which sits north of the Grand Canyon and includes the northern regions of both Coconino and Mohave Counties. Through 1986, other populations were documented on the Arizona Strip, specifically on lands controlled by the US Bureau of Land Management (BLM). Those populations were identified in House Rock Valley and along the rims of canyons tracing the Little Colorado and Colorado Rivers in Coconino County and in Hurricane Valley, in Main Street Valley, at Salaratus Draw, and on sites near Clayhole Ridge and Sunshine Ridge in Mohave County. Beyond the Arizona Strip, populations of Fickeisen plains cactus were known from the Cowpers' original site of collection near Cameron, in the Gray Mountain region of Coconino County, and from lands owned by the state of Arizona, by Kaibab National Forest, and by private citizens in areas of the county lying south of Grand Canyon National Park. Populations also were located farther east in Coconino County on lands owned by the Navajo Nation. A notably large population was reported there, specifically at Hellhole Bend, in 2009.

In total, just 33 populations of Fickeisen plains cactus, representing 1,132 individuals, were recorded between the time of Benson's studies in the

1960s and the completion of surveys conducted by BLM researchers and the Navajo Nation through 2012. As implied by their geographical distribution, those few populations occur in a disjunct fashion. Across some sites, they are separated from one another by an average distance of 19 miles.

Despite their geographical separation, every population of Fickeisen plains cactus shares in common a unique affinity for Kaibab limestone, a chert-rich layer that forms the rimrock of the Grand Canyon and the upper layer of parts of adjoining plateaus. In addition to containing microscopic quartz crystals, of which chert is comprised, the upper limestone layer also contains deposits of salt and gypsum. In many places, the limestone is overlain by a soil layer, but in other areas, such as on well-drained hills, canyon rim margins, and flat ridgetops, erosional processes have removed the soil layer above to expose the limestone and its shallow, gravelly loam soils. Fickeisen plains cactus favors such exposed sites at elevations between 4,200 and 5,950 feet.

The cactus also seems to favor places where landscape features trap moisture in the air and create pockets of rainfall. Overall precipitation on the Colorado Plateau is influenced by fluctuations in factors such as sea-surface temperature and atmospheric circulation in the Pacific Ocean. But locally, the amount of precipitation that a given area receives depends primarily on elevation and topography. On average, the region of the plateau inhabited by Fickeisen plains cactus sees between 6 and 14 inches of precipitation each year. Precipitation comes in only two periods annually: in winter and during the monsoon season in summer.

The gravelly soils and the climate in the habitat of Fickeisen plains cactus support unique desert grassland and desert scrub communities. Three distinct vegetation communities are found across its range, including the Great Basin desert grasslands, Great Basin desert scrub, and plains grasslands. Plants in those communities that associate with Fickeisen plains cactus include big sagebrush (*Artemisia tridentata*), black gramma (*Bouteloua eriopoda*), blue gramma (*B. gracilis*), brome grasses (*Bromus* spp.), broom snakeweed (*Gutierrezia sarothrae*), cat's-eye (*Cryptantha* spp.), desert-thorn (*Lycium* spp.), fourwing saltbush (*Atriplex canescens*), globemallow (*Sphaeralcea* spp.), James' galleta (*Pleuraphis jamesii*), needlegrass (*Stipa* spp.), rabbitbrush (*Chrysothamnus* spp.), rose-flowered foxtail cactus (*Coryphantha vivipara* var. *rosea*), spinystar (*Escobaria vivipara* var. *rosea*), Utah agave (*Agave utahensis*), and the cactus *Echinocactus polycephalus*.

Despite the associations, however, overall vegetation cover in Fickeisen plains cactus habitat is thin, leaving open spaces between plants. Those spaces typically are covered with biological soil crusts, which are specialized communities made up of organisms such as microfungi, cyanobacteria, brown algae, lichens, and mosses. The organisms and their by-products bind soil particles together to form the crusts, which overlie very soft soils. Biological soil crusts carry out important ecological activities, such as nitrogen and carbon fixation and soil stabilization, that support the growth and survival of surrounding vascular plants, including Fickeisen plains cactus.

The way in which Fickeisen plains cactus occupies its habitat is highly variable. In some populations, individuals are scattered widely over an area, whereas in others, they are clumped together. Population size is similarly variable, with some populations consisting of just two or three individuals, and the largest populations being made up of about twenty plants. But it is uncertain whether those numbers represent the true population size of Fickeisen plains cactus, for while it is clear that the species is rare and in decline, its exceptionally small size and its habit of retracting beneath the ground to escape cold and dry conditions make locating the plant extremely difficult. Even with information on the precise locations of populations and aided by GPS in the field, one can have a hard time relocating individual plants and even entire populations.

Fickeisen plains cactus is quite distinct in appearance, owing primarily to its small size. Mature individuals stand only 1 to 2.5 inches above the ground and measure 2 inches in diameter. The species is a succulent, globose cactus that is unbranched or occasionally branched and that lives for about ten to fifteen years. Similar to other barrel- or button-shaped cacti, the stems of Fickeisen plains cactus are covered by tapering, cone-shaped projections known as tubercles. Projecting from the tubercles are spines, the spongy, or "corky," nature of which is a distinguishing trait. In Fickeisen plains cactus, the tubercles spiral around the stem and support three to seven radial spines, which measure from more than one-tenth to just over two-tenths of an inch in length. Among the radial spines, there typically is one pale-colored central spine, characteristically bent upward, that measures about half an inch in length. The central spine distinguishes the species from its closest relative, Peebles Navajo cactus (*Pediocactus peeblesianus* var. *peeblesianus*).

Beginning around the middle of April, Fickeisen plains cactus produces flowers that often rival the size of the plant itself. The flowers are cream to yellow or yellowish green in color, with the outer tepals bearing a faint brown to purple midstripe. The inner tepals and outer tepals are inverse in shape, with the former being lanceolate, or more pointed on their upper

Figure 9. Fickeisen plains cactus (*Pediocactus peeblesianus* var. *fickeiseniae*). (Credit: Kara Rogers)

end, and the latter oblanceolate, or more rounded on their upper end. The flowers emerge from the apex of the stem and open for just one or two days, usually in midmorning.

The last flowers of the year are produced in the middle of May, around the same time that the first fruits begin to appear. The fruits are top shaped and two- to three-tenths of an inch in diameter. When they first appear, they are green and turgid, but as they mature and dry out, they turn reddish brown. After drying, they split open, the break tracing a thin vertical seam in the wall of the ovary. The seeds contained inside the fruit are small, wrinkled, and dark brown to black in color.

Once released from the fruit, the seeds presumably are dispersed in some fashion, though exactly how is unknown. Some researchers are not convinced that seed dispersal really happens at all in Fickeisen plains cactus, since seeds sometimes remain beneath or near the parent plant. There has been some speculation, however, that seeds remaining near parent plants might eventually be carried away in runoff, most likely during the summer monsoon season, which usually begins in July. Either way, dispersed or not, once the seed is set in June, the parent plant does not linger aboveground. As soon as the dry season begins to set in, typically in early June, the plants retract into the soil, either pulling themselves completely beneath the ground or sinking in until the crown is even with the soil surface. Fickeisen plants that reemerge in the fall following late summer monsoonal rains retract again in the cold winter season. Plants that do not reemerge in the fall simply remain in the soil until spring temperatures and moisture levels are just right to coax them back up. Some plants remain retracted for prolonged periods of time—three or more years in some instances.

Fickeisen plains cactus is cold adapted, and the temperature and precipitation cycles that characterize the climatic conditions of its habitat and influence its retraction are thought to also influence its flowering in the spring. The rise and fall in temperature and the concentration of precipitation events in winter and summer, in other words, ultimately encourage its reproduction. Fickeisen plains cactus is not capable of sexual reproduction until it is nearly half an inch in diameter, at which point it begins to produce flowers and fruits. Its capacity for fruit and seed production increases as the plant grows larger. Small Fickeisen plains cacti, or those less than an inch in diameter, produce anywhere from one to three fruits, whereas individuals that are about twice that size produce from two to five fruits. Most documented Fickeisen plains cacti, however, are about eight-tenths of an inch to just over an inch in diameter, and of the populations surveyed on BLM lands, little more than one-third, on average, were found to be actively reproducing.[3]

The relationship between large plant size and increased fruit production is not unique to Fickeisen plains cactus, nor is low reproductive capacity. In fact, both have been observed among other species of *Pediocactus*. And similar to Fickeisen plains cactus, other *Pediocactus* seem to lack a means of seed dispersal. The latter might explain why endemism is a common theme among species of *Pediocactus*—if seeds do not travel, populations of plants do not travel. It has been hypothesized, too, that the *Pediocactus* species that exist today are specialized relicts, possibly derived from a once-common *Pediocactus* ancestor.[4]

There are other parallels in life history among species of *Pediocactus* that may provide some insight into Fickeisen plains cactus and the mysteries of its existence. For example, reproduction among other *Pediocactus* commonly occurs via cross-pollination, in some instances involving native bee species. The flowers of the threatened Siler pincushion cactus (*P. sileri*) are regularly visited by species of *Anthophora* and *Ashmeadiella* bees. Specific pollinators of Fickeisen plains cactus have not been identified, though it is probable that cross-pollination is an important mechanism of sexual reproduction for the species. Bee flies and hover flies have been seen at its flowers.

Some *Pediocactus* species also have overlapping territories. For example, the range of Fickeisen plains cactus overlaps with that of Siler pincushion cactus and Kaibab plains cactus (*P. paradinei*), and in House Rock Valley, its range runs up against that of Brady pincushion cactus (*P. bradyi*). The potential range areas, or at least the areas of potential habitat, for several *Pediocactus*, including Fickeisen plains cactus, are thought to be much larger than what field researchers have observed. Thus, some species of *Pediocactus* may have been widely distributed historically, with their ranges having contracted in recent times.

Most *Pediocactus* species are very rare. In addition to Fickeisen plains cactus, Brady pincushion cactus, Knowlton's cactus (*P. knowltonii*), Peebles Navajo cactus, and San Rafael cactus (*P. despainii*) are listed as endangered. Threatened members include Siler pincushion cactus and Winkler cactus (*P. winkleri*). The threats facing these species are many and diverse, and they are natural and human in nature. For example, endemism likely is the result of natural processes, such as changes in local climate that occurred thousands of years ago. Hence, the species' abundance is low naturally. In the Southwest, recent long-term drought and climate change, which have been linked to human activity, probably have compounded the effects of natural processes, contributing to reductions in population size. Trampling from livestock and damage from human foot and off-road vehicle traffic are considered to be serious problems for *Pediocactus* whose habitats are threatened by livestock and human encroachment and whose diminutive size makes them especially vulnerable to loss by those means. Nonnative, invasive plant species threaten to crowd out some *Pediocactus*, and small mammals such as rodents and rabbits threaten to nibble off vital plant parts.

For Fickeisen plains cactus, it is a combination of factors and cumulative stressors that poses the most significant challenge for it and its habitat.

Climate change, livestock trampling, invasive species, off-road vehicle use, road construction, commercial development, mining operations, unauthorized collection, and rodent predation have all been examined as possible contributing factors in its decline. To know which of those factors have combined to place the greatest stress on the species, however, one must look at the overall picture of the landscape and of the factors, natural and human made, that influence it.

The Southwest is home to one of North America's most distinctive landscapes, evidenced most strikingly by its terrain, including its basins, deserts, plains, plateaus, mountains, and mesas. Every one of these features plays host to a unique group of plants and animals, such that the creatures and vegetation found in one desert area differ from those that exist in a desert in another part of the region. A major reason for those differences is adaptation, which many life-forms endemic to the southwestern United States express in obvious ways—such as the production of small, ephemeral leaves or the ability to endure extreme heat by hiding in the sand—making the region a superb laboratory for the study of life through time.

Behind many of life's adaptations in the Southwest is climate. More than ten thousand years ago, when the Laurentide Ice Sheet blanketed the greater parts of modern-day Canada and the northeastern United States, the prevailing westerly winds that now blow over the Southwest and influence its modern climate were shifted southward. In addition, the Aleutian Low, a low-pressure system located in the North Pacific Ocean and typically associated with cloudy, wet conditions, was more intense than it is today. As a result of those atmospheric conditions, the Southwest was relatively moist and cool. Snow line and timberline were several thousand feet lower in elevation than they are now. Deep freshwater lakes occupied the now-dry valleys of the Great Basin, and the overall amount of precipitation supported a relatively high groundwater table and extensive drainage networks. In the high mountains, the flora of alpine tundra and steppe communities prevailed. Woodlands covered the low mountains and areas that today are home to desert plant communities.[5]

With the retreat of the Laurentide and the subsequent rise in temperatures in the Early Holocene, moisture levels in the Southwest began to decline. Evaporation led to the shrinkage and eventual disappearance of lakes. Plant communities began to change. Some plants that had evolved during the Upper Miocene (11.6–5.3 million years ago) and that had survived in the low woodlands persisted and became members of newly formed desert communities. Woodlands in the low mountains and subalpine forest communities at intermediate elevations either persisted or migrated upward to cooler elevations. The emergence of summer monsoonal and winter precipitation, with dry periods and fires between the wet seasons, favored the evolution of vegetation communities specially equipped to deal with the climatic and ecological nuances unique to different desert, basin, and mountain settings. To those communities were added migrant plants, ones fleeing old habitats in search of conditions more amenable to their survival. By about four thousand years ago, the reshuffling process had ended, and the newly emerged plant communities in the Southwest had become firmly established. Although evolution has continued, fueling the divergence of species and the modification of communities, the species and vegetation communities that we see in the region today are a direct window to that past era of change.

Since that time, climate in the Southwest has been highly variable, being influenced largely by jet streams and atmospheric phenomena associated with the nearness of the Pacific Ocean, the Gulf of California, and the Gulf of Mexico. Since the late twentieth century in particular, the southwestern United States has experienced significant and unprecedentedly rapid changes in precipitation and temperature and in water resources. From about 1950, the region began to warm, more so than at any other time in the past six centuries. From 2001 to 2010, it saw more heat waves than any decade the century before, and streamflow reductions of 5 to 37 percent occurred in major Southwest drainage basins. Streams fed by snowmelt flowed earlier in the year.[6]

Increased heat and changes in streamflow in the Southwest have been compounded by severe drought in the late twentieth and early twenty-first centuries. In the Southwest, drought exacerbates the already restrained precipitation cycle, where the torrential downpours of the monsoon season and the snow and rain in winter have in recent centuries been separated by periods of no precipitation that may last as many as three months. Over the past five hundred to one thousand years in the Colorado River Basin, which covers nearly all of Arizona, decades-long drought has not been infrequent. In fact, based on reconstructions from tree-ring data, long and severe droughts occurred roughly one or two times each century. Although those droughts covered areas much larger than that affected by the drought of the late twentieth and early twenty-first centuries, the latter nonetheless ranked among the most severe droughts on record in recent times. And it differed in that it was thought to have stemmed largely from human-associated climate change and other human impacts on hydrology in the Southwest, rather than from natural phenomena alone.

Increases in greenhouse gases associated with human activity have been linked to unusually warm late-winter temperatures, reductions in spring snowpack, and accelerated runoff in the Southwest. Canals, dams, pumping stations, reservoirs, and other forms of water infrastructure have not only diverted water from major sources, such as the Colorado River, thereby precluding its ability to feed downriver plant and animal communities, but also separated people from water sources, such that farmers in particular have had to rely almost exclusively on groundwater resources and wells for irrigation and other forms of water use. Water diversion and the depletion of groundwater are anticipated to worsen with the increasing demands placed on water resources by ever-expanding human populations in the Southwest. As a result, in the coming decades, it is likely that drought will increase in frequency, intensity, and duration throughout the region.

The impacts of climate change on Fickeisen plains cactus are only beginning to be understood. Analyses of the reproductive effort of individual plants within populations have suggested that recent drought and unusually intense heat have resulted in decreased production of flowers and seeds. Decreased precipitation in winter might also be having harmful effects, primarily in the form of preventing seedling survival. In addition, as winters grow increasingly warmer, there could be consequences for the phenology of Fickeisen plains cactus. Rather than remaining retracted into the soil until April, for example, plants may emerge earlier, possibly leading to detrimental changes in flower and seed production or to mistimed pollination events.

So it is up against those forces of modern climate change that Fickeisen plains cactus finds itself. But to make things worse, climate change is not the only factor that challenges the species' existence. On top of the dry soil, add livestock, whose hooves tear up biological crusts and potentially dislodge plants. Livestock also compress and erode the soil, compromising its ability to store seeds and to serve as viable habitat for seedlings. Plants also may not be able to retract into compacted soil. Although there have been only a few reported instances in which livestock have negatively impacted Fickeisen plains cactus, livestock trampling disturbs desert grassland and scrub vegetation communities. The same is true for other forms of trampling, whether by human foot or vehicle.

The vegetation communities are further disturbed by the presence of nonnative, invasive species. Invasive plants often co-opt water and nutrient resources, reducing their availability to native species. They also alter pollinator behavior, such as by producing nectar that draws pollinators away from native plant species or by introducing nonnative pollinators into an ecosystem. At least fifteen nonnative, invasive plant species have been identified in the habitat of Fickeisen plains cactus on BLM lands. Additional invasive species have been reported from private lands. Examples of invasive plant species in those areas include camelthorn (*Alhagi maurorum*), cheatgrass (*Bromus tectorum*), medusahead (*Taeniatherum caput-medusae*), prickly Russian thistle (*Salsola tragus*), red brome (*B. rubens*), Russian knapweed (*Acroptilon repens*), and saltlover (*Halogeton glomeratus*). The three biggest invasive threats for Fickeisen plains cactus appear to be cheatgrass, red brome, and redstem filaree (*Erodium cicutarium*).

In general, invasive plant species occur in the vicinity of Fickeisen plains cactus at varying densities. And although it is not fully known how they might affect the cactus, we can look to the known behavior of the invasives in question to infer certain details. Red brome, for example, flowers early in the spring, taking advantage of winter precipitation and thereby posing as a source of competition for resources on which Fickeisen plains cactus may depend for flowering. Cheatgrass thrives in areas where habitat has been disturbed, making it a potential competitor for resource use in areas where, for example, livestock are present. In addition, both cheatgrass and red brome produce numerous seeds in wet years and benefit from wildfire, becoming more abundant after a burn. The greater the density of invasive species, the greater their competitive force. For Fickeisen plains cactus, increased competition for resources could mean a reduction in seedling germination, leading to low recruitment.

By becoming more abundant, cheatgrass and red brome also encourage wildfire, but in regimes that often are different from those that are native to ecosystems. Fickeisen plains cactus is thought to be adapted to frequent, low-intensity grass fires, which historically affected some areas of its range. It is not, however, adapted to the kinds of frequent, high-intensity burns that support cheatgrass and red brome. Some populations of Fickeisen plains cactus, such as those on canyon rims and ledges, are at relatively low risk of high-intensity fires. But others are in prime territory for such fires. Furthermore, there is some evidence to support the idea that wildfire and invasive species could affect Fickeisen plains cactus directly. In House Rock Valley on the Arizona Strip, cheatgrass invasion after fire is thought to have contributed to the decline of Kaibab plains cactus. Researchers suspect, too, that populations of invasive plant species will increase in the future in the Southwest, making them a serious threat to Fickeisen plains cactus and other endemic cacti and native plants in the region.

The landscape of Fickeisen plains cactus has been further disturbed by human activities such as uranium mining, road construction, off-road vehicle use, and commercial development. Because some of these activities, such as road construction, have already occurred without apparent impact on Fickeisen plains cactus, and because others are believed to be unlikely to directly impact populations of the cactus due to its tendency to cling to canyon rims and areas unfavorable to human traffic, they are not considered to be immediate threats. In areas of recreational use, such as along the Little Colorado River, features such as developed paths and paved sidewalks have kept foot traffic off occupied habitat.

Still, there are Fickeisen plains cactus populations that occur in popular tourist areas that are unprotected, and the threat of development looms on the horizon, particularly on the Navajo Nation, where a ban on development that was implemented in 1966 as a result of land disputes between Navajo and Hopi tribes was lifted in 2009. The Navajo have been considering a development project for a stretch of canyon rim along the Little Colorado River gorge that is popular with tourists. The Salt Trail Canyon Fickeisen population lies near the intended site of development.

In addition, small mammals, including rodents and rabbits, have been known to nibble on Fickeisen plains cactus. Observations have been made specifically for plants on BLM lands, but small mammal herbivory on the cactus is suspected to exist across its range. The moisture contained in the plant is especially attractive to small, thirsty animals, many of which are undaunted by the presence of cactus spines. Evidence suggests that small mammals depend more on succulent cactus in dry years than in wet years. And that may be true for Fickeisen plains cactus. In 1992 twenty-six plants were lost to small mammal herbivory at the North Canyon plot on BLM lands. That year happened to see unusually low winter precipitation. There is extensive evidence that other species of *Pediocactus*, including Brady pincushion cactus, Peebles Navajo cactus, and Knowlton's cactus, have suffered losses due to herbivory. In the case of Knowlton's cactus, increased predation occurred in years of drought.

Small mammals may also eat the plant's seeds. Rodents target fruits that have just ripened and eagerly consume the fruit, apparently leaving few seeds behind. Intense seed predation by rodents was identified as a possible cause of low recruitment in Knowlton's cactus. Whether rodents destroy *Pediocactus* seeds or scarify them for germination and thereby act as agents of seed dispersal is unclear. With the prospect of increasing drought duration and intensity, however, small mammal predation is expected to increase, likely to the detriment of Fickeisen plains cactus and its relatives.

The existing combination of threats to Fickeisen plains cactus has contributed to significant losses in numbers of reproductive adult plants and in the recruitment of young plants into populations across sites on BLM lands and on the Navajo Nation (sites where the species has been monitored more or less consistently). On the Arizona Strip, thirteen Fickeisen plains cactus populations inhabit roughly 91 acres. In 1991 those populations consisted of 323 plants. By 2012 they were down to 89 individuals. The largest study plot on BLM lands, at Dutchman Draw, where the cactus is surrounded by dense, tall grass, saw more than a 95 percent decrease in that time, having dropped from a high of 219 plants in 1992 to just 5 individuals in 2012. The decline at Dutchman Draw occurred despite the presence of plants of reproductive age.

Other populations on BLM lands have shown significant fluctuations in size, with plant numbers gradually increasing for a time before declining and, in some cases, increasing again. In general, however, most populations on BLM lands suffer from a lack of age-class diversity, being dominated by aging adult plants and having very few young individuals, indicating poor recruitment. In one population studied, an average of just two new recruits entered the population each year, far fewer than the number of dead and missing (presumably retracted) plants that had been counted.

By 2013 members of the Navajo Nation Heritage Program had made multiple visits to four of fifteen known Fickeisen populations present on their lands. Of the four populations, two were found to have suffered substantial declines, whereas the other two appeared to be stable. One of the declining populations was located at Salt Trail Canyon, where 119 plants recorded in 2006 had been reduced to 70 by 2011. The health of the remaining individuals at the site had declined markedly, too. The average diameter of plants in the population shrank from a little over an inch to a little under an inch, a change that was due not to a shift in the ratio of young plants to adult plants but to the physical shrinking of plants brought on by drought over the period from 2008 to 2010. Many individuals also apparently spent little energy on reproductive output. Some flower buds were aborted, and some flowers were produced too late in the season. Not surprisingly, fruit production was found to be exceedingly low.

Overall, populations of Fickeisen plains cactus have declined nearly 60 percent since 1992. The almost complete absence of seedling recruitment, which is vital in sustaining the long-term survival of plant populations, is especially worrying, as is knowing that climate change may be a factor contributing to the species' decline. Mitigating the impact of modern climate change requires interest and action at individual and local levels, as well as policy and regulation at national and international levels, where opposition has been intense and counterproductive. But integrating all the pieces into a workable solution is necessary if we are to slow and eventually reverse our impacts on the environment. And it may be critical for even our smallest and most prickly species, including Fickeisen plains cactus, its threatened and endangered *Pediocactus* relatives, and acuña cactus.

It is remarkable to think that in 1975, when officials identified Fickeisen plains cactus and acuña cactus as candidates for the endangered and threatened list, the concept of climate change was unfamiliar to most people. It is even more amazing that the two species were initially identified as candidates in the absence of any suggestion that climate change was a threat to their populations. They were simply rare, and their rarity, endemic nature, and small population sizes warranted further assessment in order to determine their status.

Acuña cactus is a similarly fascinating plant. It, too, is a spherical cactus, but it is much larger than Fickeisen plains cactus, measuring, at its largest, about 16 inches in height and 3 or 4 inches in width. On its tubercles, it has between eleven and fifteen 1-inch-long radial spines and three to four central spines. Similar to Fickeisen plains cactus, its central spines bend upward. They differ, however, in being not only greater in number but also longer, about 1.5 inches in length, and mauve in color. Acuña cactus flowers are pink to purple and about 2 to 3 inches in diameter. They characteristically emerge at the top of the plant, several in number, and they are pollinated by small, native bees, primarily a species of solitary cactus bee (*Diadasia rinconis*) and a species of leafcutter bee (*Megachile palmensis*).

The fruits of acuña cactus are small, about half an inch in length, and pale green. They ripen in April, owing to the cactus's distribution in the warm, southerly region of Arizona and in the Mexican state of Sonora. Acuña cactus grows on low knolls and gravel ridges and in valleys, being far less choosy about the bedrock type that it settles on than Fickeisen plains cactus. It does, however, seem to prefer to grow beneath the canopy of other plants, perhaps taking shelter from temperature extremes. Its plant associates are characteristic of the Sonoran Desert and include catclaw acacia (*Acacia greggii*), creosote bush (*Larrea tridentata*), desert ironwood (*Olneya tesota*), palo verde (*Cercidium microphyllum*, or *Parkinsonia microphyllum*), and triangle bur ragweed (*Ambrosia deltoidea*).

Several populations of acuña cactus, including one at Organ Pipe Cactus National Monument, have been monitored since the 1970s. From 1981 to 2011 the Organ Pipe population declined from an estimated 10,000 individuals to 1,000 to 2,000 plants. On BLM lands that house the species,

Figure 10. Acuña cactus (*Echinomastus erectocentrus* var. *acunensis*), Little Ajo Mountains near New Cornelia Mine, Pima County, Arizona. (Photo credit: US Fish and Wildlife Service)

some 310 living plants counted in 1987 had been reduced to 75 plants by 2006, with no juvenile plants seen from 2008 to 2013. Comparable losses were documented on both state-owned and privately owned lands.

Some of the factors that threaten Fickeisen plains cactus have also been examined for their impacts on acuña cactus. For example, livestock grazing, invasive species, animal predation, drought, and climate change all could pose some risk to acuña cactus. Other potential threats include urban development, which tends to ebb and flow with the housing market in southern Arizona, and border activity, a threat wholly unique to those portions of the southwestern US states that lie adjacent to Mexico.

The land on which acuña cactus has made its home is some of the most unforgiving in the Southwest, owing not only to the terrain, heat, and dryness but also to the nature of human activity in the region. Some of the privately owned lands where acuña cactus occurs are isolated, largely abandoned, or uninhabited properties. Some have been for sale for years, and they have been treated as wastelands by humankind.

Stronger than that sense of isolation and abandonment is the feeling of trespass that permeates the land. As a protected and relatively isolated area, Organ Pipe Cactus National Monument holds special appeal for those seeking to illegally cross into the United States from Mexico. Illegal drug and human smuggling activity occurs daily in the park, where there have been violent run-ins with cross-border violators in the past. More than three-quarters of remaining acuña cactus plants are in either Organ Pipe or adjacent Sonora, Mexico, within about 10 miles of the border. In Organ Pipe, the plants are in an area that has been closed to visitors specifically because of illegal border activity.

In 2006 the United States completed work on a fence that runs along the border region of the park. The fence has helped to reduce illegal vehicle traffic in the area, but new trails and roads were created as a result of increased border enforcement. Those changes added to the thousands of miles of land in Organ Pipe that were already affected by unauthorized roads and trails. And while those changes resulted in an apparent drop in violators, thousands of immigrants continue to attempt to cross the border each year. Many of those people do not travel on existing paths. Their movement is unpredictable, constantly changing in order to evade border agents. Furthermore, patterns in movement have been influenced by the installation of communication towers in and close to Organ Pipe. No towers were built near acuña cactus plants, but the shift in human traffic patterns may have encouraged people to travel closer to or into acuña habitat. Tire tracks and footprints have been found in acuña cactus study plots, suggesting that there is real potential for direct mortality of plants from those activities. Vehicular and foot traffic can also disturb the plants that associate with the cactus, cause erosion or soil compaction, and facilitate the introduction of seeds and plant material from invasive species.

In addition, plants in the border region but lying outside protected lands have been threatened by the harvesting of desert ironwood and mesquite (*Prosopis velutina*). In northern Mexico, the two species have been targeted as a source of fuel for various activities and as a wood resource for craftmaking in the tourism industry. The bare land left behind has been converted for livestock grazing, representing a significant change for acuña habitat.

Acuña cactus also is susceptible to insect predation, particularly by the cactus longhorn beetle (*Moneilema gigas*) and by cactus weevils (*Gerstaeckeria* spp.). Cactus longhorn beetles are large, flightless beetles whose larvae bore into the stems and roots of cacti, potentially severing those parts and killing the plants. Adults can cause damage to plants by feeding on pads and terminal buds. Adult cactus weevils similarly feed on surface structures, but their larvae bore into pads and stems to feed internally, eventually killing infested structures. Although other insects are known to impact acuña cactus, cactus longhorn beetles and cactus weevils are thought to be behind the decline of certain acuña populations. Plants that have been found toppled over in the Coffeepot Mountain acuña population, for example, may have been attacked by cactus longhorn beetles. In other instances, plants have been found uprooted, suggesting that javelinas or possibly birds may have dug them up. Small mammals, such as pack rats and mice, have been known to feed on acuña cactus as well, leading to the potential loss of plants.

Other factors that have been linked with mortality and low recruitment in acuña cactus are drought and climate change. Annual rainfall has been shown to be positively associated with the number of flowers that individual acuña plants produce.[7] In Organ Pipe, acuña cactus is one of the first plants to flower in the spring, suggesting that winter precipitation may be especially important for flowering. Several studies support that idea, having found correlations between the number of flowers produced and the amount of precipitation received the preceding winter. Winter precipitation may also affect fruit set. In the summer, monsoon precipitation and soil moisture appear to be vital for seedling survival.

Low or absent recruitment in many acuña populations has been coincident with drought in the Southwest. Continued summer and winter drought stress, in tandem with warmer winters in the region, could mean that few acuña populations will see new individuals cropping up in their ranks.

Those conditions could also mean prolonged breeding cycles for the plant's insect predators. Stressed from drought and heat, acuña cactus is less able to recover from the mortality that results from insect attacks. Normally, the production of seedlings would provide some measure of protection, but decreases in reproduction and seedling recruitment associated with drought and climate change are thought to have severely compromised the survival of acuña cactus.

Because Fickeisen plains cactus and acuña cactus are newly listed as threatened, plans for their recovery are only beginning to be developed. It is clear that recovery will not happen quickly for either species, given the complexity of factors that threaten their existence. Human-associated climate change and drought, for example, are long-term issues, adequate solutions for which may take years to implement effectively.

Under the Endangered Species Act, the two species and their habitats are protected. Some 47,123 acres of land have been proposed for designation as critical habitat for Fickeisen plains cactus.[8] The acreage is divided across federal, state, tribal, and private lands, though the majority—28,300 acres—falls on federal and state property. For acuña cactus, proposed critical habitat covers 18,921 acres, the majority of which is again made up of federal and state lands. If those critical habitat areas are accepted and given adequate protection, then both Fickeisen plains cactus and acuña cactus would at least be guaranteed a home.

With suitable habitat available, one could speculate that seedling production and planting programs, which would help to increase populations within native habitat areas, may eventually be established for both cacti. However, cultivating either species may not be a simple matter. One issue concerns the collection of seeds from plants in the wild. In the coming years, wild seed sources for cultivation of either species are likely to become increasingly sparse. So long as drought conditions persist, plants will continue to experience poor flower production and seed set. Seeds from cultivated plants could be genetically assessed for their similarity to wild plants and used as a last resort for recovery programs.

In the case of acuña cactus, a seed bank already exists. However, long-term viability of seeds is poor, based on greenhouse trials in which between 0 and less than 19 percent of seeds that were four to six years old germinated successfully. Seeds that are one to two years old appear to have better germination rates. Fickeisen plains cactus also has been cultivated successfully by growers, and plants are sold by commercial vendors. Seeds are also available for sale. The genetic resemblance of those seeds to wild specimens, however, would need to be evaluated before they could be used for species recovery. Furthermore, even with seeds from cultivated plants, Fickeisen plains cactus can be very difficult to grow, in part because of its susceptibility to diseases such as root rot. Whether cultivated individuals of either species would survive transplantation into the wild is another matter altogether.

Other components of recovery for Fickeisen plains cactus and acuña cactus could include measures to minimize small mammal and insect depredation and measures to address the impacts of development. For Fickeisen plains cactus, the control of invasive species will be important, and continued study to determine the extent to which activities such as livestock grazing and off-road vehicle use impact the cactus will be needed. For acuña cactus, ongoing monitoring of the impacts of border activity will be critical, as will be investigation of animal activity.

Of the nineteen cactus species that were federally listed as endangered as of 2013 in the United States, seven were endemic to parts of Arizona. Others were located in surrounding southwestern states, including California, Colorado, New Mexico, Texas, and Utah. A couple others were native to Florida. Too often the environments inhabited by cactus are viewed as wastelands, largely unsuited for human settlement and considered to be unproductive for agricultural, business, or commercial development. But biologically and ecologically, the places that cactus species call home are among the most diverse and complex in North America. And they are beautiful.

In the Southwest, we share with plants and animals a powerful connection to those places: the need for water. Every living thing in the desert is thirsty. For humans, extreme thirst sets in quickly, and our ability to survive without water lasts but a brief time. But species of native animals and plants, including the tiniest members of the region's flora, can persist. They may suffer, but as species, they survive. And they have done so for thousands of years, often with little or no protection from the sun, the heat, and the relative lack of water. Knowing that makes our native species of cactus all the more amazing, and it makes their protection all the more necessary and meaningful.

Penstemon

With more than 270 species, the genus *Penstemon* is one of North America's largest groups of plants. It is also one of its most enjoyable. From their beautiful flowers to their relative ease of cultivation and extreme tolerance of drought, penstemons have long been a source of interest and delight for plant enthusiasts. They have also been of great importance in the medicinal and cultural traditions of Native Americans.

But ages before Europeans described them and Native Americans integrated them into their lives, penstemons simply were a part of the North American landscape. They continue to be a part of that landscape today in a wild existence that is the most captivating of all their qualities. One can encounter penstemons in the wild in almost any region of the North American continent, from one coast to the other and across a variety of habitats, including wetlands, meadows, and deserts. In all those places, penstemons hold in store little surprises, such as an unexpected floral hue or a liking for an unusually hostile soil environment. Their diversity is striking. Their flowers span the spectrums of color, shape, and size, and their differences in biology, ecology, and geography are unlike those of any other group of plants in North America. Importantly, their remarkable depth of distinction and their unexpected character have kept researchers asking questions, even now, centuries after the genus was first recognized botanically. The result is an ever-expanding body of botanical and horticultural knowledge of penstemons.

From a formal scientific perspective, the history of penstemons began in 1741, when botanist and physician John Mitchell introduced the genus name. *Penstemon* means "five stamens," in reference to the number of those structures possessed by each plant. Of the five stamens, however, one is merely stamenoid (sterile) and without a recognizable anther. Mitchell probably used the name *Penstemon* to describe either of two species found in the eastern United States, hairy beardtongue (*P. hirsutus*) or eastern smooth beardtongue (*P. laevigatus*).[1] He published the name in 1748.

Soon after Mitchell's introduction, the genus name became a source of great confusion. Plants assigned to *Chelone*, the establishment of which predates that of *Penstemon*, also possess five stamens, the fifth one being sterile. So, despite Mitchell's introduction of *Penstemon* as a separate genus, the pioneer of binomial nomenclature, Carolus Linnaeus, disregarded it and placed the plant Mitchell described within *Chelone* in keeping with previous supposed relationships for the group. He also used an alternate spelling, *pentstemon* (with *pente-*, meaning "five"), to refer to a plant known as *Chelone pentstemon* (later called *Penstemon pentstemon*).

Confusion over the taxonomic position and spelling of *Penstemon* persisted. But by the late eighteenth century, at least a few botanists had begun to see what Mitchell had seen: a group of plants that warranted a category all their own. Increasing numbers of botanists treated the *Penstemon* genus as separate from *Chelone*, and Mitchell's original spelling was increasingly accepted. Later, the taxonomic separation was supported by the identification of key differences in flower and seed morphology. In *Chelone*, for example, the inflorescence is simple, whereas in *Penstemon* it is always compound. The opening of the corolla in *Chelone* is covered by an anterior lip, whereas in *Penstemon* it is covered only in rare instances, and then by an arching anterior lip. Another key difference is in their seeds: those of *Chelone* are winged, those of *Penstemon* are not.

The taxonomic history and subtle morphological differences may seem academic, but they underlie all that is relevant to modern horticultural and botanical understanding of *Penstemon*. The separation of *Penstemon* and *Chelone* made way for the expansion of the former. At the beginning of the nineteenth century, fewer than half a dozen penstemons had been described. By 1818, however, the genus had more than doubled in size, thanks to the publication of *The Genera of North American Plants* by Thomas Nuttall, who had the wherewithal to classify his new finds according to Mitchell's arrangement. Of special note from *The Genera* was Nuttall's summary of *Penstemon*: "A North American genus and probably an extensive one."[2] Not long after Nuttall's work was released, Scottish botanist David Douglas, during his explorations of western North America, discovered another eighteen penstemon species. Douglas had explored only the

Pacific Northwest, which left a lot of land to be covered in the search for penstemons new to botany.

In a way, fulfilling Nuttall's prophecy of "an extensive" genus not only in number but also in geographical distribution, explorers and researchers who later took to characterizing the life of the western plateaus and intermountain regions found still more penstemons. Species were discovered in southern deserts, including those in Mexico, and in high alpine environments, such as those in Colorado. Remarkably, the discovery of several penstemon species has been very recent, occurring within just the last few decades.

In North America, the recognition of *Penstemon* as a distinct group of plants has been important for our knowledge of the continent's native flora and for our understanding of plants and plant ecology. It not only resulted in the expansion of *Penstemon* into our largest group of native plants but also helped botanists to appreciate the remarkable extent of plant and habitat diversity on the continent. Penstemons became a powerful lens by which scientists are able to explore the effects on plants of a wide range of environmental factors.

Individual penstemon species are not especially abundant in the wild. Rather, they tend to occur in small populations and often are limited in distribution. In some cases, that limitation is extreme, restricting plants to a single type of substrate or to the vicinity of a single stream, for example. Because historical records are lacking for many penstemons, it is uncertain whether the different species ever thrived in any great number or broad distribution.

For many penstemons, the extremes in habitat restriction and their significance for each species' persistence in the presence of threatening factors were identified only recently. Several penstemon species that are proposed as threatened also were discovered only recently, after factors such as climate change and oil production were identified as potential threats to other species now known to share penstemon habitat. Some of those other species, similar to the penstemons they live alongside, are endemic to the places they inhabit.

As of 2013 in the United States, two species of penstemon were listed as endangered: blowout penstemon (*P. haydenii*) and Penland beardtongue (*P. penlandii*). Parachute penstemon (*P. debilis*) was listed as threatened, and Graham beardtongue (*P. grahamii*) and White River beardtongue (*P. scariosus albifluvis*) were designated as proposed threatened. Although other penstemon species were not recognized federally as threatened or endangered, many were recognized as such by individual states. For example, in

Washington, Barrett's beardtongue (*P. barrettiae*) was listed as threatened. The same designation was given to lilac beardtongue (*P. gracilis*) in Iowa, while in Michigan that same species was considered to be endangered. In New Jersey and Ohio, eastern smooth beardtongue, one of the earliest described penstemons, was listed as endangered. Still other species were recognized as being of special concern or were given special protection in various US states.

But of all known penstemons and, by some accounts, of all known plants in North America, the threatened Parachute penstemon is thought to be the most rare. It also embodies the exceptional struggle for protection that confronts penstemons.

Parachute penstemon is a relatively recent addition to the *Penstemon* genus, having been discovered in the mid-1980s by Steve L. O'Kane, Jr., and John L. Anderson.[3] O'Kane and Anderson came across the unfamiliar plant while conducting research on the status of four other species of plants endemic to the Piceance Basin in northwestern Colorado. Within the Piceance Basin lies the Roan Plateau, a roughly 230-square-mile area of flat uplands carved by deep canyons at its southern end. The Roan Plateau houses the remaining populations of Parachute penstemon, which occur on steep, south-facing talus slopes between 5,600 and 9,250 feet in elevation. The pitch tumbles down dramatically, in some places giving way to high cliffs. The perilous slopes are well known to those who live in or who have passed through Parachute, Colorado, which lies a short distance from the penstemon sites.

Parachute penstemon is a diminutive plant, about 2 inches in height, that has devised a clever means of holding itself in place while gravity works on the loose rocks around it. Part of its strategy involves growth in the formation of mats, in which fleshy roots and stems arise from an underground shoot system that extends from the plant's woody base. The subterranean shoots sometimes root at their nodes, helping anchor plants into the ground on the unstable, down-shifting talus. Providing additional stability are its buried stems, which though weak (hence *debilis*) emerge from the shoots and elongate in a down-slope direction in search of surfaces stable enough to support them. The growth habit of Parachute penstemon is similar to that of Red Canyon penstemon (*P. bracteatus*), which is endemic to an area of limestone scree in the vicinity of Bryce Canyon and Red Canyon in south-central Utah, and to that of Harbor's beardtongue (*P. harbourii*), which is endemic to talus and scree slopes in Colorado's high alpine.

The blue- or gray-tinted leaves of Parachute penstemon are oval-shaped, thick, and succulent. They are positioned oppositely on the stem, with each

Figure 11. Parachute penstemon (*Penstemon debilis*). (Credit: Kara Rogers)

leaf measuring just under an inch in length and a half-inch in width. In June or July, white to lavender flowers bloom, funnel-form and small, with the corolla reaching a length of about seven-tenths of an inch. Similar to other penstemons, the corolla possesses two distinguishable lips, the lower being slightly longer than the upper.

The anthers of Parachute penstemon are magenta at the time of flowering. Whether that is of any consequence for pollination is unclear, but Parachute penstemon is visited by various common species of pollinators when its flowers are in bloom. Although no pollinators have been found to specialize exclusively on Parachute penstemon, native bees appear to do the bulk of its pollinating. Following cross-pollination, fruits develop, typically appearing from mid-July to August and containing tiny seeds that measure less than a tenth of an inch in length. Seeds are dispersed by gravity. They simply fall from the parent plant and nestle into the soil below. Despite the requirement for cross-pollination, populations of Parachute penstemon may be affected by inbreeding depression, as there is limited genetic diversity within and limited contact between populations.

Considering that its habitat is characterized by a shifting substrate that lacks surface soil, it is amazing that Parachute penstemon seedlings are able to take root at all, much less thrive and become reproductively active. The soil, which might be more accurately described as rubble, is part clay and part shale fragments. It is dry as well. The region where Parachute penstemon occurs technically is a high desert, receiving fewer than 13 inches of precipitation annually.

Its demand for that specific substrate composition and climate may explain why it is found only on certain talus slopes on the Roan Plateau. Alternatively, its tolerance of the soil and the shifting talus environment may give it an edge over dominant plant species in the area, allowing it to avoid competition and thereby survive, albeit in an unlikely setting. The choice habitat for Parachute penstemon supports few other plants. Among the species' scattered plant associates are four species endemic to the area, including Arapien blazingstar (*Mentzelia argillosa*), Cathedral Bluff meadow-rue (*Thalictrum heliophilum*), dragon milkvetch (*Astragalus lutosus*), and oil shale fescue (*Festuca dasyclada*). Nonendemic associates include Colorado bedstraw (*Galium coloradoense*), Henderson's wavewing (*Pteryxia hendersonii*), mountain mahogany (*Cercocarpus montanus*), mountain monardella (*Monardella odoratissima*), rayless tansyaster (*Machaeranthera grindelioides*), rockspirea (*Holodiscus dumosus*), spearleaf buckwheat (*Eriogonum lonchophyllum*), and yellow rabbitbrush (*Chrysothamnus viscidiflorus*).

The shale habitat of Parachute penstemon is generally unfavorable for plants, because it is harsh, requiring that vegetation find unusual solutions to survival, such as a subterranean, down-slope growth habit. The shale substrate is also geologically ancient, having been derived from the Parachute Creek Member, a geologic subdivision of rock in the Green River Formation on the Roan Plateau. The Green River Formation is dated to the early to middle portion of the Eocene epoch (56–33.9 million years ago). Within the formation, the Parachute Creek Member is one of the richest oil shale deposits in the United States. By some estimates, it has the potential to yield billions of barrels of oil.

From the mid-twentieth century until the first part of the twenty-first century, research into oil shale development in the Piceance Basin intensified due primarily to growing interest in domestic fuel production, which proponents argued would decrease US reliance on petroleum from foreign countries. At the same time, however, researchers also discovered a variety of previously unknown plants endemic to the basin. That is what drew O'Kane and Anderson to the Piceance Basin in the mid-1980s, when they discovered Parachute penstemon. They were concerned specifically with the status of Cathedral Bluff meadow-rue, Dudley Bluffs bladderpod (*Physaria congesta*), Dudley Bluffs twinpod (*P. obcordata*), and Piceance bladderpod (*P. parviflora*). All four of these endemics were first collected or described in the late 1970s and early 1980s, and all are rare.

The work of O'Kane and Anderson, as well as that of other researchers, uncovered a common theme for native vegetation in the Piceance Basin: endemism. And not unlike certain other species that share its habitat, Parachute penstemon is more specifically an oil shale endemic. It may even be a neo-endemic of that type, a species recently evolved and now reproductively isolated in its habitat because of its dependency on or ability to tolerate shale. Other genera of oil shale endemic plants, including species of *Astragalus* and *Physaria*, are thought to be rapidly evolving and perhaps neo-endemic as well due to the close proximity of relatives, the inherent genetic variability of the species, and the many different features of the terrain and sedimentary strata in their habitats, which provide homes for newly emerging forms.

The idea that Parachute penstemon may be neo-endemic is supported by phylogenetic evidence, in which the whole of the *Penstemon* genus was found to have experienced a recent continental radiation, with new species emerging in unique ecological niches due mainly to evolutionary adaptation.[4] The finding makes the process of continental radiation fundamental to our understanding of how the different species of penstemons, including Parachute penstemon, ended up where they did. The process of radiation is also linked to the diversity of species within *Penstemon*, owing to evolutionary adaptation and the development of unique traits, such as specialization to a certain type of pollinator.

Biogeographically, the continental radiation of *Penstemon* began in the Rocky Mountains, a conclusion based largely on analyses of the tribe Cheloneae in the family Plantaginaceae (to both of which *Penstemon* belongs).[5] Investigations have also indicated that the distribution of early members of Cheloneae extended from the Neotropics, across the Bering land bridge, and into extreme eastern Asia. Probably in the middle of the Tertiary Period, the eastern Asian species diverged from their North American counterparts, which themselves split into western and eastern groups. Five of the seven genera contained within Cheloneae, including the genus *Penstemon*, share an overlapping distribution in the Cascades and Sierra Nevada. The same region also happens to be a major center of diversity within the *Penstemon* subgenus *Dasanthera*, which is the basal lineage of the genus.[6] Another center of diversity within *Dasanthera* is the northern Rocky Mountains. Based on leaf surface traits and biochemical characteristics, those penstemon species classified in subgenus *Dasanthera* that occur in the northern Rocky Mountain region appear to be the more primitive of the two diversity centers.

Their primitive nature is further supported by evidence from genetic and biogeographic studies of *Dasanthera*.

From the Rocky Mountains, *Penstemon* likely migrated to the Cascades–Sierra Nevada ranges, and from there some species went west, while others radiated across the Intermountain Region and into the Southwest. Eventually, the genus crossed over the Rockies and spread onto the Great Plains before finally expanding into eastern North America. The events underlying the continental radiation of *Penstemon* are thought to have coincided with events of the Pleistocene epoch (2.6 million–11,700 years ago).

The spread of *Penstemon* eastward beyond the foothills of the Rockies ultimately also landed Parachute penstemon in oil shale habitat, though not directly. Parachute penstemon is classified in the subgenus *Habroanthus*, and it therefore is not the most primitive of penstemons. Thus, a species ancestral to Parachute penstemon likely entered the region first and then, through specialization to different habitat types, split into new forms. That hypothesis is based in part on the general pattern of continental radiation for the genus, as well as on the idea that Parachute penstemon is a neo-endemic species, which would make it relatively recent in origin.

Little is known about the actual distribution of Parachute penstemon, other than what information is available for the described populations on the Roan Plateau. The total area occupied by those populations amounts to roughly 2 by 17 miles, which is a remarkably small amount of land, considering that it houses all the individuals for the species. Previous estimates of total geographical range were even smaller. Prior to 1995, only two populations of Parachute penstemon were known—one immediately below the peak of Mount Callahan, and the other on Roan Cliffs at Anvil Points Rim. A survey conducted in 1996 resulted in the identification of only two new occurrences, both of which were located near to the two previously described populations. Later surveys recorded the existence of several other occurrences, as well as the disappearance of almost the entire Anvil Points population.

By 2011 Parachute penstemon existed at just seven locations: Mount Callahan Natural Area, Mount Callahan Saddle Natural Area, Anvil Points Rim, Anvil Points Mine, Mount Logan Road, Mount Logan Mine, and Smith Gulch.[7] Viability at those locations ranged from poor at Mount Logan Road and Anvil Points Rim to excellent at Mount Callahan Natural Area, with the others ranking as fair or good. The two populations that received rankings of poor viability were later deemed nonviable, given that they were disjunct and exceedingly small. The occurrence at Mount Callahan Natural Area was by far the largest, housing around 2,200 individuals within an area of 32.7 acres. By comparison, the nonviable population at Anvil Points Rim was the smallest, with just two plants having been recorded within a 5.7-acre area in a follow-up survey published between 2008 and 2011. The total number of Parachute penstemon plants remaining in the wild was estimated in 2011 to be about 4,138 individuals.[8]

Several populations appeared to be trending slightly downward in size. An exception was the Anvil Points Rim population, which experienced a drastic decline. Where 2 plants stood in 2011, 250 had been recorded years earlier. As suggested by their viability rankings, both the Anvil Points Rim and the Mount Logan Road populations, the two smallest populations remaining, were on the brink of extirpation. They were, in all likelihood, beyond hope of recovery. Their situation did not bode well for other small populations, particularly the one at Smith Gulch, where about fifty plants were left.

Other occurrences of Parachute penstemon may exist, since some potential habitat areas with dangerously steep terrain have been left largely unexplored, and other potential sites occur on private lands where surveyors were not granted access previously. However, the prospects of those areas having viable populations of the species seem slim, if the recent extensive surveys, which were carried out in the most promising locations but which turned up only a couple occurrences, are any indication. Of note, that trend, in which only a few additional plants were found, diverges from trends among several other plant species endemic to the Parachute Creek Member, where the species were found to be more common than thought after further surveys were carried out.

A significant factor working against the survival of Parachute penstemon is that more than 80 percent of all remaining plants, as well as about two-thirds of the species' habitat, reside on lands owned by companies that specialize in energy development. The rest of the plants and habitat occur on land leased from the Bureau of Land Management (BLM) for the exploration and development of oil and gas resources. So the greatest threat to Parachute penstemon is energy development, which includes oil development, gas development, oil shale extraction, mine reclamation, and the construction and maintenance of roads.[9] Each of these activities brings with it an array of known and potential environmental impacts, the nature and extent of which depend on the technologies employed and on how lands are accessed and maintained.

Natural gas extraction can happen by several different means, depending on the geologic source. Conventional natural gas reservoirs are places where free gas has become trapped in a porous zone of a rock formation, typically either carbonate, sandstone, or siltstone. To meet rising natural gas demands, however, companies have increasingly turned to unconventional natural gas reservoirs, which include tight (largely impermeable) coal seams, shales, and sandstones. The shift to unconventional reservoirs has brought with it a move to relatively hazardous extraction processes. In the case of conventional natural gas, once a potential source is located, a well is drilled to allow the natural gas to be lifted from its underground reservoir. By contrast, in order to release gas from unconventional reservoirs with any degree of efficiency, deep, often horizontal boreholes are required, through which a fracturing fluid that consists of water, sand, and chemicals is injected under pressure. The latter step, known as hydraulic fracturing (or fracking), creates fissures that enable the release of trapped gases and the passage of those gases to the wellhead.

The chemicals used in hydraulic fracturing, however, as well as the high-pressure injection methods employed, introduce certain risks for the environment. For example, the use of fracturing chemicals raises the potential for the contamination of sources of surface water and groundwater. Whether the chemicals actually make it into groundwater is a topic of much debate, but they have been detected in surface water near and downstream of drilling sites. Furthermore, research has shown that certain chemicals used in natural gas drilling trigger estrogen and androgen receptor activity in animals, which has the potential to lead to abnormalities in development, immune function, neurological activity, and reproduction.[10] Plants, in helping to absorb or clear chemicals from the environment, may experience detrimental effects, too.

The impact of fracturing chemicals in surface water and groundwater samples on hormone receptor activities has been examined specifically in Garfield County, Colorado, which houses the Piceance Basin.[11] Water samples were collected from sites within the Colorado River Drainage Basin and the Piceance Basin that had been fractured for natural gas extraction, had experienced a natural gas spill or drilling-related incident (e.g., produced water tank leak or gas upwelling), and housed from about 40 to more than 130 natural gas wells within a 1-mile radius of the sampling site. Of the samples tested, estrogenic activity, antiestrogenic activity, androgenic activity, and antiandrogenic activity levels exceeded levels measured at reference control sites. Among the chemicals analyzed in the study were benzene, ethylene glycol, naphthalene, and styrene, which have been shown to negatively impact plants.[12]

In addition to the study's findings, two other important details accompanied the research. One was that more than 100 of the 750 chemicals used for hydraulic fracturing are known or suspected endocrine disruptors. The other was that natural gas spills are common in the area. Hundreds apparently occur in Garfield County each year. Furthermore, hydraulic fracturing is associated with other environmental concerns, among the most significant of which is that the injection of fluids to release gases has the potential to increase pressure on seismic faults, making them more likely to slip and thereby inducing earthquakes. Induced seismic activity, similar to water contamination, is a hotly debated issue that requires further investigation to establish a causal link.

Oil shale mining and processing is likewise a highly involved endeavor, complete with an extensive list of potential avenues of environmental harm. Because oil substances are present in solid form in oil shale, which prevents them from being pumped up from the ground, oil shale mining is inherently more complex than conventional oil recovery. To get at the oil in oil shale, companies have experimented primarily with two different approaches: so-called surface, or ex-situ, retorting and in-situ retorting. Surface retorting involves mining oil shale and bringing it to the surface, where liquid oil is extracted by heating the oil shale to high temperatures. In in-situ retorting, the oil shale is heated underground, and the resulting liquid is pumped to the surface.

The two approaches have different but equally significant impacts on the environment. Surface retorting has the potential to radically transform a landscape because of the opening of mines and the deposition of spent shale on the surface, typically either in a surface impoundment or as fill, which is sometimes dumped back into a mine. In addition, although areas that contain spent shale deposits may be reclaimed, several characteristics of spent shale challenge that effort. For example, spent shale is salty and basic, and it is dark in color, causing it to absorb large amounts of heat. It also contains residual amounts of chemicals used in processing, among them hydrocarbons. As a result, spent shale is not amenable to plant growth. Plants that survive the high salinity and the chemical-infused substrate could be affected by extreme heat. The heat can become so intense on south-facing slopes that it can kill seedlings that attempt to root there. These problems may be overcome when spent shale is buried beneath a couple feet of soil, but soil compaction and erosion have the potential to limit the growth of plants that require deep rooting systems. Furthermore, the chemicals in spent shale can also be a source of surface water or groundwater pollution, leaching from the shale over time. Leaching is a significant problem for in-situ retorting, where the retort process is more difficult to control beneath the ground.

Oil and gas exploration and development, as well as the development of oil shale extraction techniques, have been ongoing in the Piceance Basin since about the middle of the twentieth century. Development and exploration have occurred in fits and starts due to limitations and advances in technology and fluctuations in markets. In-situ retorting, for example, remained a largely experimental technique in the first decade of the twenty-first century, and oil shale production had yet to reach commercial levels in the United States by that time. Similarly, the use of hydraulic fracturing in the region had remained relatively limited.

Natural gas production in the basin, on the other hand, was nearing all-time highs in 2013, and hundreds of drilling permits were awarded annually to companies in Garfield County. Although only a portion of those permits were actually put into action, the decades-long time frame and the cumulative extent of drilling in the basin suggested that environmental disturbance was well under way there. Exactly how much that disturbance had affected Parachute penstemon was unclear, but surface disturbance, contaminated runoff, altered runoff flows, dust generation, waste production, mine closure activity, and inadvertent trampling by humans were either potential or demonstrated threats.[13]

In addition to the surface disposal of wastes such as spent shale, surface disturbance can occur in a variety of other forms within Parachute penstemon habitat. For instance, road construction, off-road travel, pipeline construction, the development of well drilling pads, and the creation of evaporation ponds alter surface environments. They can also affect patterns of water runoff, lead to trampling, and result in the generation of dust and waste. Chemical spills or spills of other wastes, as well as chemical leaching from spent shale, can lead to the contamination of runoff and groundwater. Many of the aforementioned activities can also result in a loss of habitat for both Parachute penstemon and the insect species that pollinate it.

In the case of road construction, new or widened roads mean an inevitable increase in vehicular traffic. Roads in Parachute penstemon habitat are especially problematic, as they tend to cut across talus slopes, making them susceptible to mudslides and rockslides. They therefore require significant maintenance, which for Parachute penstemon may be just as damaging as, or even more so in the long run, the construction of the road itself. At Anvil Points Mine, for example, road maintenance claimed the lives of several plants, and road widening at Mount Logan Mine left plants dangling by their roots.

Vehicular traffic also raises the risk of dust deposition. Though the actual extent of harm to Parachute penstemon from dust is unknown, over time dust could be a serious threat for the species. According to estimates by the US Forest Service, a vehicle that travels 1 mile of unpaved roadway once a day every day for a year causes 1 ton of dust to be deposited along a 1,000-foot-wide corridor that extends from either side of the median.[14] For any plant, too much dust can clog stomata, thereby impairing gas and water exchange, raising leaf temperature, and leading to deficiencies in photosynthesis. The overall result is a reduction in growth and vigor. Several populations of Parachute penstemon lie within about 300 feet of unpaved roads, placing them in prime dust deposition territory. Those populations also have little to no protection in the way of barriers to shield them from dust. As a consequence, the potential for dust deposition may be quite high in some areas of the species' habitat. One such site is the Mount Logan Road population, which lies close to a road frequently traveled by trucks.

Impacts from dust deposition on Parachute penstemon could work in concert with other threats within the shale habitat and could threaten other plant species that share space with Parachute penstemon. Declines in multiple plant species could result in a loss of pollinator habitat and pollinators. The loss of pollinators would translate into fewer pollination opportunities for Parachute penstemon.

Parachute penstemon's relationship with its pollinating insects is already in jeopardy due to the small size of most of its populations. Small plant populations (those with fifty or fewer plants) offer limited nectar rewards for pollinators. As a result, they tend to receive fewer visits from pollinators compared with relatively large plant populations. With few plants and limited seed set, genetic diversity within a small population is poised to decline. Such a decline appears to be under way for Parachute penstemon, according to genetic analyses and comparisons with plants with life histories like its own.

In recent decades, Parachute penstemon's populations have been reduced in size primarily through division and destruction by habitat fragmentation associated with human activity. Plants have also been lost, however, as a result of activities that ultimately are intended to be beneficial. In removing health and safety hazards at Anvil Points Mine, which included the closure of entry points into the mine and the elimination of soil contaminated by lead, seventeen plants were killed. Another thirty died at Mount Logan Mine in a reclamation project that involved the addition of topsoil, beneath which the plants were buried. Prior to mining and reclamation, the Parachute penstemon population at Mount Logan may have been fairly large, suggesting that the population that exists there now is a remnant of the former one. If activities in the area continue as they have, the current population could become extinct.

In addition to energy development activities, natural phenomena, human-associated climate change, and drought also threaten Parachute penstemon. Rockslides, for example, could abolish small populations, while competition with other native plants could expedite the loss of small populations. At Anvil Points Rim, competition with native yellow rabbitbrush is thought to have facilitated the demise of indigenous Parachute penstemon plants, all of which disappeared in the late 1990s. Transplanted individuals were introduced to the site in the mid-1990s, but nearly all of them had died by 2010 for reasons that were unclear. Erosional processes within an ephemeral streambed at the Smith Gulch location represented another natural phenomenon that jeopardized the persistence of plants.

Although not studied extensively for Parachute penstemon, climate change and drought could also have consequences for the species. Its restriction to oil shale habitat might mean that it would be slow to adapt to other substrates, potentially hindering its migration to escape warming temperatures. In such a situation, the species would most likely attempt to migrate vertically where possible.

The threats to Parachute penstemon are not going away anytime soon, and in fact those that have already affected the species will probably worsen, and potential threats will become actual threats. That prediction is based simply on general trends in human population growth and increasing energy demands within the United States and abroad. Fulfilling those demands will be challenging, and the United States is expected to shift increasingly to natural gas and alternative energy sources. The country is also likely to be an increasingly important player in energy exports.

With so much at stake, it is simple to see why those who are invested in energy development might not give much or any thought to a 2-inch-tall plant clinging to an existence in one of North America's oil-rich regions. So the question becomes how best to protect Parachute penstemon when the vast majority of its habitat lies on lands owned by energy development companies. Some of those companies have denied access to researchers and state and federal biologists who were seeking to discover new populations of Parachute penstemon or to simply monitor existing ones.

The majority of those companies probably will never engage in conservation agreements, but they do have a conservation model to follow. Though it affords protection for only about 40 percent of the species' habitat, the model in question includes almost 70 percent of plants, representing the two largest Parachute penstemon populations: Mount Callahan and Mount Callahan Saddle Natural Area.[15] The lands on which those two populations occurred were owned by Occidental Petroleum Corporation (Oxy USA

Inc.). In 1987 Oxy agreed to help protect Parachute penstemon by engaging in a voluntary and nonbinding arrangement with the Colorado Natural Areas Program (CNAP). The initial agreement included just one Parachute penstemon population, but in 2008 it was expanded to include the second Mount Callahan population. In 2012 it was expanded again to include the Mount Logan Mine area.

As part of the agreement, Oxy implemented best-management practices that were intended to reduce threats to and impacts on the species. Ultimately, the strategies included the establishment of surface disturbance buffer zones, the management of noxious weeds, the introduction of limits on motorized travel, the protection of pollinators and pollinator habitat, and the control of dust. Impacts on the species also were reduced by considering conservation needs when developing new drilling pads. With the help of funding provided by Oxy, CNAP officials monitored the species and tracked population trends.

Being voluntary in nature, the agreement was based on trust, but it had survived for decades. And it did so despite the fact that the efforts by both parties might not ultimately prevent the species from gaining an endangered listing. That upgrade in risk could happen when energy demands rise, since increased demand for energy would be accompanied by increased drilling activity, which would escalate the impacts on Parachute penstemon. Although greater demand for energy might be beneficial financially for Oxy, having an endangered species on its lands is less than ideal, given the federal involvement that comes with an endangered listing. The company's efforts to help conserve Parachute penstemon were some of the longest in action for the species and among the more effective to have been implemented. Oxy announced in 2013 that it had authorized select asset sales, which included certain midcontinent assets, among them its holdings in the Piceance Basin. The potential sale of those assets made the future management of Parachute penstemon in affected areas uncertain.

On BLM lands, efforts to manage Parachute penstemon habitat have included the monitoring of plants, the closure of mine tunnels, and the transplantation of individual plants at risk from mine reclamation activities. In addition, in 2008 the Anvil Points region was designated an Area of Critical Environmental Concern (ACEC). The Anvil Points ACEC includes the Anvil Points Mine population and the Anvil Points Rim habitat. Though the lands were available for lease, the stipulations of the ACEC included no surface occupancy and no ground disturbance in the Anvil Points area, with certain caveats.[16]

In 2012 the federal government published its final rule on critical habitat for Parachute penstemon. Critical habitat was defined by the following features: the presence of Parachute Creek Member outcrops with steep slopes, the presence of a Rocky Mountain Cliff and Canyon plant community with minimal plant cover (less than 10 percent), an elevation between roughly 6,300 and 9,300 feet, a high-desert climate that receives between 12 and 18 inches of annual precipitation, and suitable pollinators and pollinator habitat.[17] Four critical habitat units were identified that met those specifications. They included Mount Callahan, Anvil Points, Brush Mountain, and Cow Ridge. The latter two units were not occupied by the species, though they included areas yet to be surveyed at the time of the rule. In addition, since they met the critical habitat criteria, they were considered to be ideal areas for future recovery and introduction projects for Parachute penstemon. The region of Mount Callahan that was included in the designation consisted of land where the species occurred primarily as spillover from the two major populations on the mountain, the bulk of which otherwise lay on Oxy property. The four areas totaled 15,510 acres. Total suitable habitat for the species, which included occupied habitat, was estimated at 16,862 acres.[18]

Successful recovery of Parachute penstemon within its critical habitat will occur only when existing and introduced populations are stabilized and pollinator habitat of sufficient area is maintained. To reach these goals, specific actions, which involve primarily surveying and monitoring, threat abatement, and research, must be carried out. Surveys are especially important in the effort to identify new populations, particularly at Brush Mountain and Cow Ridge. Monitoring will allow for the elucidation of population trends and will be vital in areas affected by reclamation activities. Threat abatement encompasses a number of different aspects, including the prioritization of sites for protection and the implementation of protection. The latter can be achieved through efforts to prevent leasing and energy development in critical areas and coordination with landowners and companies to limit impacts on Parachute penstemon, particularly impacts sustained through road construction or maintenance. Threat abatement also entails land acquisition, namely of private lands that house the species, as well as the establishment of permanent conservation easements.

The third specific action, ongoing research on Parachute penstemon, was expected to provide further insight into the species' ecology, pollination, and propagation. Additional knowledge about the species that pollinate Parachute penstemon plants could help with the establishment of buffer zones, which may be seeded with other native plants that share the same pollinators or that provide habitat for Parachute penstemon's pollinators. Likewise, knowledge of which plants may compete with Parachute penstemon is needed to ensure that buffer zone species do not encroach upon the penstemon's habitat. Propagation studies could result in the identification of techniques and habitat conditions that facilitate the establishment of seedlings and transplants on federal lands.

Conservation practices, regulations, and agreements that strike a balance between species protection and energy development are critical. Parachute penstemon is not the only penstemon species, nor the only plant species for that matter, whose federal protection has been overshadowed and hindered by energy interests in western North America. Of particular concern are Graham beardtongue and White River beardtongue, which by the end of 2013 were still listed as proposed threatened species. Both species are found only in northeastern Utah's Uinta Basin and the neighboring area of Colorado, where they inhabit calcareous soils derived from oil shale. The soils are associated with the Green River Formation and more specifically with the Parachute Creek Member. Despite their limited population sizes and ongoing threats, which include oil shale, natural gas, and tar sands development, neither species had gained a federal listing sufficient to ensure their long-term protection. Similar to Parachute penstemon, energy development interests have more or less dictated their fate.

There is, across the Green River Formation, much at stake in terms of North America's natural heritage. What we stand to lose is evident from what exists in Colorado's Piceance Basin alone. In the 1970s and 1980s the BLM carried out a series of environmental assessments for a proposed oil shale leasing program.[19] The goal was to provide an environmental impact statement that the secretary of the interior could use in guiding decisions about whether to lease BLM land in the basin. The assessment examined the impacts of oil and gas development on a range of factors in the region, including air quality, water quality and hydrology, climate, soil, plants and animals (including humans), cultural resources, geologic and paleontological resources, and aesthetics. It also provided a summary of the basin's living and nonliving resources.

The assessment indicated that the Piceance Basin is home to large animals such as mountain lions, mule deer, and, in rough winters, elk. Raptors, such as golden eagles, marsh hawks, red-tailed hawks, and rough-legged hawks, are also present in the basin in different seasons. Owls likely nest in tracts of pinyon-juniper woodlands in the area. In addition, several cultural sites were found to be eligible for inclusion in the National Register of Historic Places, and four geologic units, representing important sites of fossil discovery, were recorded. Much of the unique plant life in the basin, however, was yet to be documented, and for all that was found to exist in the Piceance Basin, the BLM leasing program moved forward.

Despite the political weight of its primary source of threats, there is hope for Parachute penstemon. Awareness and concern for its future are growing. Its future will be determined largely by the individuals and companies based in the Piceance Basin. If they fail as stewards of nature, they will be held responsible for the loss of a species and a part of America's natural history. That is an immense responsibility to carry, one that many people might not shoulder gracefully or willingly. But to those who endeavor to protect plants and nature, the burden of responsibility is an opportunity, a chance to ensure a future for Parachute penstemon.

Golden Paintbrush

Each spring, the few tranquil meadows and prairies to be found on the islands of Puget Sound and Canada's Vancouver Island come alive with color. In some places, purple blends with green to accentuate otherwise subtle dips and rises in the landscape. In other places, yellow brightens the fields by degrees, as though competing in brilliance with the sun above and heralding the arrival of warm weather.

Yellow is a particularly attractive addition to the landscape of the Puget Sound region and nearby Canadian islands, in part because it occurs in different ways—a patch here or there on a rocky hillside or the better part of a small meadow aglow. Some of these areas are graced by the nestled heads of buttercups or by the drooping flowers of yellow bell. In certain meadows on a small collection of islands in the region, magnificent vertical strokes of yellow blaze and bring the imagination to life.

The first image that comes to mind is one of little yellow rockets, their tips flaming as they stretch up from the earth below. Another invokes a picture of flowers that resemble upside-down paintbrushes dripping with yellow or, more precisely, gold. Both descriptions are fitting, and, more importantly, they are applicable to only one species in the world, golden paintbrush (*Castilleja levisecta*). Golden paintbrush is the only species of Indian paintbrush (genus *Castilleja*) to produce such a glorious yellow blaze, and it is the only flowering species to produce that visual effect in the meadows and prairies of northwestern Washington and the adjoining region of southwestern British Columbia, Canada.[1]

That glorious blaze, unfortunately, is now but an intimation of what it once was in that part of the world. Historically, golden paintbrush was known from more than thirty sites that existed in southwestern British Columbia and the northwestern United States, but by 2010 just eleven populations remained. In 1997 the species was listed as threatened in the United States, and in 2000 it was given an endangered listing in Canada. Washington State also considers the species to be endangered and has given it an S1 designation, meaning that it is critically imperiled.[2] In Oregon, the species was last seen in 1938, and it is now ranked as SH, or historic, since its existence there is known only from historical reports, with all naturally occurring populations believed to have been extirpated from the state.

Eight of the eleven naturally occurring populations that were left in 2010 were concentrated on Washington's Whidbey Island and San Juan Islands. Whidbey Island hosted five of those populations, which were located specifically at Admiralty Inlet Natural Area Preserve (Naas Natural Area Preserve), Ebey's Landing, Forbes Point, Fort Casey State Park, and West Beach. The three populations in the San Juan Islands included those at False Bay, Long Island, and San Juan Valley. Two populations remained in Canada, one each on Trial Island and Alpha Islet, two small islands situated off the eastern end of Vancouver Island. Just one population was left on the Washington mainland. It was located in Rocky Prairie Natural Area Preserve in Thurston County, just south and east of the Olympic Mountains. Rocky Prairie was the largest of all the remaining populations, housing just over 7,000 plants as of 2005.[3] Trial Island hosted the next largest population, 3,192 plants, counted in 2006. The smallest population was that at Admiralty Inlet, where just 94 plants were counted that same year.

The numbers are much higher than those we've seen for most species explored in earlier chapters, which partly explains why golden paintbrush is listed as threatened and not endangered in the United States. But that does not mean that the species is safe from an endangered listing, especially since its habitat continues to disappear. In Canada, golden paintbrush habitat has become severely constricted. Between Trial Island and Alpha Islet, which are very small islands, it is possible that as few as 70 acres, or just one-tenth of a square mile, of the species' habitat remain. Those islands are protected as Canadian Ecological Reserves, but they offer far less area for population expansion than the species' primary historical home, Vancouver Island, once did.

In the United States, golden paintbrush habitat has undergone drastic reductions as well, but the greater numbers of plants and populations provide at least some sense of security. That could of course be a false sense, for there is a certain degree of unpredictability about golden paintbrush. To begin with, published population counts for the species are old, and the data that are published show, for example, that at Ebey's Landing, the population count dropped from 669 in 2005 to 214 in 2006, whereas at Forbes Point, the count jumped from 123 to 260 over that same period. And while ten-year and five-year trends suggest that some populations have stabilized and may even be increasing, others are in decline. For still others, data are insufficient to make predictions about their future. Add to that the unexpected loss in 2000 of a fourth population that had once existed at Davis Point (Lopez Island) in the San Juan Islands. That population is now extirpated for reasons that remain unclear.

With uncertainties about the viability of remaining habitat and about unpredictable fluctuations in population size, the future of golden paintbrush is insecure. Its chances of recovery appear to be high, thanks largely to its high levels of seed production and seed viability, which facilitate direct seeding into restored habitat as well as nursery propagation. Nursery plants have been successfully transplanted into reintroduction sites, and seeds sown into prepared habitat have met with some success in the production of viable seedlings.

Nonetheless, significant challenges lie ahead. Low rates of seed germination and seedling survival in the wild, for example, suggest that golden paintbrush is years away from persisting in the wild without the aid of humans. Its recovery is further complicated by the generally poor ability of measures such as genetic diversity, population size, and geographical distance to predict the performance of golden paintbrush at reintroduction sites. Normally, those measures can be used to direct reintroduction efforts because they allow conservation specialists to select seeds for planting from the most suitable populations.

On top of these issues are the constant threats posed by invasive species and habitat loss. The history of these threats with golden paintbrush is long, and much like Mead's milkweed or Florida torreya, the presence of threatening factors in golden paintbrush habitat predates scientific knowledge of the species. Importantly, however, it does not predate the existence of Native Americans in the northwestern United States or of First Nations peoples in British Columbia. The native peoples knew their regions' prairie habitats well, and their activities, specifically the setting of fire to native prairies, helped maintain the habitat. So to find the first page in the story of the decline of golden paintbrush, we turn again to the history of European American settlement in North America.

For several centuries, the lands of Oregon's Willamette Valley, Washington's Puget Sound region, and Canada's Vancouver Island were known only from the exploration of the Pacific Coast. Even at the end of the 1700s, when trade was in full swing on the coast, the only formal establishments that existed along the northern coast were trading posts. Opportunities to achieve more were seemingly ample, and yet Europeans remained within the vicinity of their few outposts. One reason for that may have been Native Americans and First Nations peoples, who had effectively repelled the foreign visitors from their territories on several occasions. But more significant, perhaps, was the question of who owned the land, which was a subject of dispute between Spain and Great Britain.

Eventually, in 1805, European Americans penetrated into the Oregon region. But they did so over land, having come from the east, for that was the year that Lewis and Clark arrived at the mouth of the Columbia River. Their expedition opened the way for settlement of the Oregon Country, an expanse that included the lands that encompass modern-day British Columbia, Idaho, Oregon, Washington, northwestern Montana, and western Wyoming. In 1843, following the breaking of the Oregon Trail, European American settlers moved into the area that now comprises Washington State. Many settlers moved specifically into the Puget Sound area, likely because of its proximity to the Pacific Ocean, which offered opportunities for work, and because of the easily traveled corridor north from Oregon, where the low-lying lands between the Olympic Mountains to the west and the Cascade Range to the east converged. The same year that settlers made their way into Washington, the first British settlement, Fort Albert (later Fort Victoria), was established on Vancouver Island.

Despite their relatively late settlement, Oregon, the Puget Sound region, and Vancouver Island underwent rapid and extensive development. The three regions attracted significant numbers of people, many of whom were drawn by the promise of work in the timber, mining, and railroad industries. At times in the nineteenth and early twentieth centuries, those industries sustained sizeable workforces. But they also were susceptible to boom or bust cycles. When work opportunities were few, economies destabilized, and workers looked elsewhere for employment. Although the ebb and flow of the economy drove some people away, as mining and timber work dropped off, people increasingly found ways to make their livings from other industries, including agriculture, tourism, and technology.

In Oregon's Willamette Valley, agriculture was, by the mid-nineteenth century, already an important economic driver. The crops and livestock raised there fed hungry miners and railroad workers who lived and worked in other parts of the state. Today, the valley is known particularly for its production of nursery plants, fruits, vegetables, and wine. It also has become the most densely populated region of the state, housing about 70 percent of Oregon residents.

On Vancouver Island, by comparison, land for agricultural endeavors was limited. In the north and west, forestry remained a viable industry for some time, having declined only recently in association with global drops in demand for pulp and other forest products. In the south and east, the economy became increasingly diverse, with technology and tourism forming key components. The result today is an island with a dense, thriving urban population on one half and a declining rural population on the other half.

The Puget Sound region was similarly affected by limitations in land available for agriculture. It also historically trended toward industries such as smelting and shipbuilding and readily adapted to the emergence of new industries, such as the aerospace industry, which became prominent in the region during the First and Second World Wars. Today, the economic contributions of several of those older industries have been bolstered by growth in the business, health, maritime, technology, tourism, and transportation sectors.

How that progress in development relates to golden paintbrush is complex, but the impacts of development have become increasingly apparent in ecosystems across the three regions. In southeastern Vancouver Island, for example, less than 5 percent of Garry oak (*Quercus garryana*) and associated ecosystems exist in a natural state. Their drastic reduction has come as a result of land conversion for development related to agriculture, commerce, housing, small manufacture, and transportation. Development has led to both the loss and the fragmentation of habitat for native species on the island, and it has increased opportunities for the introduction of nonnative, invasive species. In the Willamette Valley, native oak woodlands and savannas have been reduced to 8 percent of their historic area, while just 1 percent of the valley's wet prairie habitat remains intact. Land use and development have been the primary agents of habitat loss and fragmentation there.

There is little known about how abundant golden paintbrush once was or how extensively it was distributed historically. Much development was already under way in the Willamette Valley, in Puget Sound, and on Vancouver Island by the time American botanists came to learn of the species' existence. Among the first people to collect it were American botanists Louis F. Henderson and Thomas Jefferson Howell. The men found the plant in 1879 and 1880, respectively, both in Clark County, Washington, in the southwestern part of the state.[4] While the exact site of Henderson's find is unclear, Howell found the species near Mill Plain, which falls within the northern extent of the Willamette Valley. The first to describe the species was Jesse More Greenman, in 1898.[5] Between the time of Henderson's and Howell's collections and Greenman's description, the plant was documented at locations in the Puget Sound region and in British Columbia. One of the first to identify it in Canada was Irish-born Canadian botanist John Macoun, who collected a specimen in May 1887. British botanist Charles F. Newcombe collected it from Vancouver Island in 1890 in habitat with dry ground located near the sea.

In 1905 golden paintbrush was collected at Bonneville in Multnomah County, Oregon, which sits at the northern edge of the state on the Columbia River. Bonneville is near the Willamette Valley, and it is perhaps best known as the home of the Bonneville Lock and Dam, which was completed in 1938. It might be merely coincidence that the dam's completion occurred in the same year as the last sighting of the species in the state, but it is likely that its construction resulted in the destruction of habitat for golden paintbrush. In 1910 American botanist Morton E. Peck collected several golden paintbrush specimens from Marion County, located a short distance south of Portland. The collection sites were characterized as damp and open. Golden paintbrush subsequently was collected from multiple sites in Linn County, which also happens to be the home of the last sighting of golden paintbrush in Oregon. The date of that last sighting was May 1938, and the location was described as Peterson Butte Cemetery, or what later became known as Sand Ridge Cemetery.[6] The location of that final Oregon description for the species is characterized by well-drained, gravelly soils, similar to those found at historical collection sites in Washington and on Vancouver Island.

Greenman described golden paintbrush as "many stemmed from a perennial base." The "many" stems range in number from five to fifteen. They are unbranched, typically semierect to spreading, and clustered or creeping. Individual plants can grow to more than a foot or even a foot and a half in height, though most are shorter. Leaves occur in an alternate pattern and progress from a long, narrow, and pointed architecture at the lower part of the stem to a broad, egg-shape form at the upper part of the stem. Upper leaves often have one to three pairs of lateral lobes, whereas lower leaves are entire. The aboveground portion of golden paintbrush is covered in dense, soft hairs, which usually are sticky, particularly on the upper surfaces of leaves.

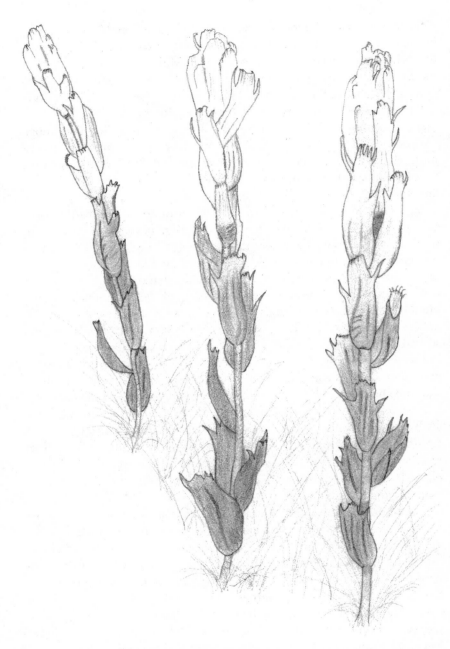

Figure 12. Golden paintbrush (*Castilleja levisecta*). (Credit: Kara Rogers)

When not in bloom, golden paintbrush is a rather inconspicuous plant. And if it were not for its floral bracts, it would be hard to notice even when flowering, given that its flowers are pale green in color. The corolla, which features fused petals, a long beak-like upper lip, and a shorter lower lip, measures just over three-quarters of an inch in length. It is hidden entirely by the plant's large, overlapping, golden to yellow bracts, which give golden paintbrush the magnificent color for which it is known. The bracts are comparable in width to the plant's upper leaves, and their tips are cut by one to three pairs of narrow lobes. Similar to the rest of the plant, the bracts are covered in sticky hairs. Flowers bloom from April to June.

While the pollination biology of golden paintbrush has not been described in detail, it is thought that bees serve as its primary pollinators. Bumblebees (*Bombus* spp.), for example, are regularly observed visiting the plant's flowers. Golden paintbrush appears to be almost entirely self-incompatible. In general, crosses between nonsibling plants from the same population or between plants from different populations produce higher seed set than self-crossed or sibling-crossed plants.[7] A key characteristic of between-population crosses is that they show no decline in seed set. Hence, individuals from different populations are the best candidates for cross-pollination. The production of seeds is thought to be vital for golden paintbrush, since the species has never been observed to reproduce by vegetative means.

Green fruits may appear as early as May or as late as July and thereafter ripen and remain on the plant. By mid-July, or once fruits have matured and summer drought has set in, the plants fall dormant. In the following weeks, the fruit capsules dry and begin to dehisce, breaking open and exposing their many minute seeds. Dehiscence usually occurs in the latter part of the summer, with some fruits remaining closed into August or September. After the fruits open, the seeds are dispersed gradually over the next several months, apparently being shaken from their parent plants by the wind and likely being blown a short distance away. Parent plants typically are still dormant during the seed-dispersal period. They emerge from their quiescent state only when soil moisture levels rise sufficiently to support the production of new growth. For most plants, that has occurred by January, at which time new shoots, generally in the form of leaf masses located at the bottom of the plant, begin to grow. Shoots elongate in March and April, ready to support newly emerging buds.

What happens once the seeds of golden paintbrush are dispersed remains somewhat open to speculation. The seeds probably do not travel far from their parent plants, which suggests that most seeds start out on suitable habitat, increasing their chances for successful germination. But as nursery and garden trials have shown, much more than a suitable bed is needed. Other conditions must be fulfilled in order for germination to occur, but precisely what those conditions are remains unknown. Factors thought to be important include seed age, with first-year seeds showing increased viability compared with second- or third-year seeds, and seed exposure to cool temperatures. Observations of seeds collected from wild plants and seeds sown under greenhouse conditions indicate that germination rates among cultivated golden paintbrush plants can vary markedly—from 39 to 96 percent, by some estimates.[8] By comparison, germination rates are much lower for wild golden paintbrush and for seeds sown in restored habitat. In addition, the viability of seeds has been found to differ for different populations of golden paintbrush. Those seeds that do survive may go on to produce plants that live for as many as five or six years.

The survival of golden paintbrush at various stages of its life is influenced by several factors. For example, similar to other species of Indian paintbrush, golden paintbrush is a hemiparasite, combining photosynthesis with parasitism. It contains all the necessary biomolecules and machinery for photosynthesis but draws water and mineral nutrients from the vascular systems of other plants via a specialized structure known as a haustorium. The haustorium is a modified root that is capable of forming a parasitic connection on the root of another plant. A number of other species of the broomrape family (Orobanchaceae), to which golden paintbrush belongs, are also hemiparasitic. Examples in North America include species of bird's-beak (*Cordylanthus*), false foxglove (*Aureolaria*), and lousewort (*Pedicularis*).

Golden paintbrush is considered to be a facultative hemiparasite, in that it does not necessarily depend on a host for its survival. Greenhouse studies have shown that golden paintbrush plants thrive when given a host but can germinate, develop, and grow in the absence of a host.[9] Whether that is also the case in the wild is unclear, though field experiments have suggested the same. For example, first-year golden paintbrush plants, when faced with mortality from vole tunneling associated with the presence of the host plant common woolly sunflower (*Eriophyllum lanatum*), experienced improved survival in experimental plots where the host was excluded. Other experiments have shown, however, that golden paintbrush plants grown with the host species Roemer's fescue (*Festuca roemerii*) experienced better survival than plants grown without a host. Furthermore, although the association is weak, in the greenhouse, golden paintbrush achieves a larger size and flowers more often when associated with a host plant than when grown without a host.

As a facultative hemiparasite, or essentially a generalist feeder, golden paintbrush can make use of different host species. For golden paintbrush, hosts identified in greenhouse studies include, in addition to common woolly sunflower and Roemer's fescue, species of checkerblooms (*Sidalcea* spp.) and slender cinquefoil (*Potentilla gracilis*). According to greenhouse and outplanting trials, however, although golden paintbrush can benefit from interactions with any of those hosts, it thrives to different degrees. And while the identification of hosts has improved researchers' ability to cultivate golden paintbrush, the role of host species in the survival of golden paintbrush in the wild is as yet unclear.

Research on other hemiparasitic plants has suggested that parasitism on leguminous hosts, which have the ability to fix nitrogen, improves the performance of the parasitic species. That could be true for species of Indian paintbrush. Entireleaf Indian paintbrush (*Castilleja indivisa*) plants that parasitize leguminous lupines, for example, produce more flowers, have higher numbers of visits from pollinators, have more pollinator probes per visit, and produce more seeds than plants that parasitize grass.[10] The nitrogen boost received from the host plant is thought to be responsible for the benefits reaped by the paintbrush plants. Lupine hosts also contain defensive compounds, primarily alkaloids, which can protect plants from herbivory and might offer secondary benefits to parasitic paintbrush plants.

In the wild, the survival of golden paintbrush is further influenced by its habitat. The species occurs in open, relatively flat grasslands or on steep, grass-covered coastal bluffs. It favors acidic, well-drained soils of glacial origin. In the northern extent of its range, including historical sites on Vancouver Island and extant sites on Alpha Islet, the Trial Islands, and the San Juan Islands, those soils are derived primarily from clay-rich glaciolacustrine sediments, which were deposited in ancient lakes by glacial meltwater. In the region of Thurston County, where the species survives at the Rocky Prairie preserve, the preferred soil is characterized as gravelly glacial outwash or coarse, heterogeneous glacial drift sediment. In the southern extent of its historic range (the Willamette Valley), the species grew primarily in clay-based, sandy alluvial soils, which were deposited by flowing water, often in floodplains. The soils that supported golden paintbrush there probably were limited to upland prairies, which would have been well drained and dominated by grasses associated with oak savanna.

Physiographically, golden paintbrush habitat lies within the Puget Trough Physiographic Province and the Willamette Valley Physiographic Province of the United States and the Georgia Depression ecoprovince in British Columbia. At the northern end of the Puget Trough lies the Puget

Lowlands, which effectively includes the San Juan Islands and adjacent Trial Islands and Alpha Islet, though the islands can be said, too, to lie within the Georgia Depression. The southern end of the Puget Trough connects with the Willamette Valley province, where the low-lying lands between Pacific mountain ranges run for another 130 miles. Glacial water in both provinces, as well as in the Georgia Depression, shaped the habitat of golden paintbrush. The Willamette Valley province, for example, essentially is an alluvial plain. It is thought to have experienced massive floods during warm, interglacial periods, which likely resulted in the deposition of glacially derived alluvial sediments throughout much of the province. But whereas upland prairies in the region tend to be covered by sandy, well-drained soils, the other predominant habitat type in the Willamette Valley province, wetland prairie, is low-lying and poorly drained. The lowlands of the Puget Trough, by contrast, are relatively well drained. They run north into the fjords of Puget Sound, revealing throughout their length the effects of glacial processes. The result of these glacial processes, which carved and excavated the lowlands between the Olympic Mountains and the Cascade Range, was the deposition of glacial outwash and sediments and the formation of basins, channels, and freshwater lakes. Water that flowed through the southern end of the trough probably contributed to flooding in the Columbia River and adjacent portions of the Willamette Valley.

The Willamette Valley, the Puget Trough, and the Georgia Depression ecoprovince are further characterized by their temperate maritime climates. Winters are mild and wet, and summers are cool and dry. In the Puget Trough, for example, January temperatures hover in the high thirties or low forties (°F), while July temperatures often are in the low to mid-seventies. Precipitation typically is in the form of rain, with areas of the Puget Trough that lie outside the rain shadow cast by the Olympic Mountains seeing anywhere from 35 to 50 inches each year. The Willamette Valley receives similar amounts of precipitation. Conditions on the islands at the northern end of the range of golden paintbrush vary with respect to overall figures for the Puget Trough. On Whidbey Island, for example, little more than 20 inches of rain falls each year, and summer highs struggle to reach 70°F.

Whether on the islands in the north or in the Puget Trough or the Willamette Valley, the vast majority of precipitation comes in the winter months. Again, though, the islands experience important variations. Soil moisture levels on the islands typically begin to rise in late summer and early fall, when dew increases as coastal fogs move in. Fog is common along the Pacific Coast in cooler months, because coastal waters generally are much cooler than offshore waters. Plants often benefit from the additional moisture supplied by dew, and that appears to be the case for the island populations of golden paintbrush. Heavy dew can cause the plant to emerge from dormancy as early as mid-September. Early emergence does not appear to negatively affect the species, even though it can leave individual plants susceptible to winter frosts. Because annual precipitation for coastal populations tends to be lower than for the population at Rocky Prairie, the dew likely brings much-needed moisture to coastal golden paintbrush plants.

Variations in soil and climate influence not only the occurrence of golden paintbrush but also that of the plant species that associate with it, whether as hosts or as other members of the vegetation community. Plant species known to share habitat with golden paintbrush include blue wildrye (*Elymus glaucus*), checker lily (*Fritillaria affinis*), common lomatium (*Lomatium utriculatum*), common woolly sunflower, common yarrow (*Achillea millefolium*), field chickweed (*Cerastium arvense*), hollyleaved barberry (*Mahonia aquifolium*), Idaho fescue (*Festuca idahoensis*), red fescue (*F. rubra*), Roemer's fescue, Nootka rose (*Rosa nutkana*), Pacific woodrush (*Luzula comosa*), small camas (*Camassia quamash*), and vetch (*Vicia* spp.).

Several of these species are found only to the west of the Rocky Mountains, and not all are distributed equally across the range of golden paintbrush. The plant species observed in golden paintbrush restoration sites in Oregon, in other words, differ from those observed in its habitat in the Puget Trough or on Trial Island. The different native plant associates also vary in their abundance. Many appear to occur only in small populations. For example, at Rocky Prairie, certain native perennials, such as checker lily, Nootka rose, and common yarrow, make up less than 7 percent of the vegetation cover. Red fescue, on the other hand, contributes to more than twice that amount.[11]

Trees are infrequent features within golden paintbrush habitat, but they form the basis of the ecosystems that support its native plant communities. Across the historical and extant range of golden paintbrush, which includes all sites in the Georgia Depression ecoprovince and in the Puget Trough and Willamette Valley provinces, the vegetation community that defines the species' habitat is associated with the Garry oak ecosystem. Garry oak, also known as Oregon white oak, is the only native oak species found in those parts of North America, and its ecosystem is one characterized as grassy and savanna-like. Its soils typically are gravelly and associated with glacial outwash.

Garry oak ecosystems are early successional, meaning that they become established relatively quickly following disturbance. That means, too, that in the absence of disturbance, they are readily overtaken by later successional species, such as the woody species of Douglas fir ecosystems. In general, the plant species that characterize early-successional ecosystems are relatively short-lived, rapidly consume soil nutrients, and have increased rates of photosynthesis and primary productivity. Early-successional plants are also highly competitive, relying on various mechanisms to outgrow and to suppress the growth of succeeding species. Some of these characteristics are evident in golden paintbrush, being reflected particularly in its hemiparasitic lifestyle and the superior viability of first-year seeds. By comparison, the species of Douglas fir ecosystems are relatively long-lived, slow-growing, and resilient. Once established, they are difficult to eliminate with the sorts of natural disturbances that benefit golden paintbrush and its Garry oak associates.

The primary type of disturbance that Garry oak ecosystems have depended on historically is fire. Up until about the mid-1800s, low- to mid-intensity fires in Garry oak grasslands were ignited naturally, such as by lightning, or were set by humans. Native Americans and First Nations peoples burned the grasslands to encourage the growth of plants such as camas, which was an important food source. Following the displacement of the native peoples by incoming European American settlers and the ensuing development of the land, however, fire suppression became the norm. The loss of fire facilitated the encroachment into native grasslands of woody species, particularly Douglas fir (*Pseudotsuga menziesii*), California blackberry (*Rubus ursinus*), snowberry (*Symphoricarpos albus*), western brackenfern (*Pteridium aquilinum*), and wild rose (*Rosa*). The intrusion of those species, although native to the region, has resulted in the degradation of native grassland habitat. As a consequence, species such as golden paintbrush, which depends on the openness of grassy living spaces, have been slowly choked out by overshading and the accumulation of forest litter.

In the southern Puget Trough and the Willamette Valley in particular, the combination of fire exclusion and urban and agricultural development spurred a dramatic decline in the areal extent of native grasslands. In the southern portion of the Puget Sound region, just 3 percent of native grasslands remain. While small and not able to host large human populations, Trial Island and Alpha Islet have nonetheless suffered damage to their native Garry oak ecosystems and grasslands as a result of settlement.[12] Aggravating the situation for golden paintbrush in all those locations has been the invasion of nonnative species, which have severely altered species composition in what few grasslands remain.

Increasingly, the plants found in golden paintbrush habitat have been identified as nonnative. For example, in the Willamette Valley, among fifteen plant species that were identified as indicators of golden paintbrush habitat, just two were native species, while at sites in the Puget Trough, about half of indicator species were native.[13] In the Puget Trough, significant invaders included nonnative Kentucky bluegrass (*Poa pratensis*) and the introduced species narrowleaf plantain (*Plantago lanceolata*). In the Willamette Valley, a variety of nonnative species were found across sites. Among them were black medick (*Medicago lupulina*), blue fieldmadder (*Sherardia arvensis*), brome fescue (*Vulpia bromoides*), bull thistle (*Cirsium vulgare*), changing forget-me-not (*Myosotis discolor*), silver hairgrass (*Aira caryophyllea*), sticky chickweed (*Cerastium glomeratum*), tall oatgrass (*Arrhenatherum elatius*), and Queen Anne's lace (*Daucus carota*). In Canada, several exotic species have been identified in the maritime meadow habitat of golden paintbrush. Frequently occurring exotics there include gorse (*Ulex europaeus*) and Scotch broom (*Cytisus scoparius*). Nonnative English ivy (*Hedera helix*) and the naturalized species English holly (*Ilex aquifolium*) and Himalayan blackberry (*Rubus armeniacus*) are growing threats in Canada's maritime meadow habitat.

Nonnative species tend to be precocious, taking up residence across a wide geographical range. For that reason, some introduced species are found in golden paintbrush habitat areas in both Washington and Oregon and on the Canadian islands and islands in the Puget Sound region. Common velvetgrass (*Holcus lanatus*), Kentucky bluegrass, and orchardgrass (*Dactylis glomerata*) are a few examples of nonnative species found in both the Puget Trough and the Willamette Valley. Gorse and Scotch broom are common in golden paintbrush habitat on the Canadian islands and at Ebey's Landing on Whidbey Island, for example. In addition, similar to native plant species, nonnative species vary in their abundance in golden paintbrush habitat.

Nonnative plants are problematic for golden paintbrush in that they compete for vital resources and sometimes grow with remarkable vigor in response to disturbances, potentially allowing them to rapidly take over large swaths of golden paintbrush habitat. Nonnative Scotch broom, for example, because it germinates quickly after fire, is particularly menacing for golden paintbrush recovery strategies that involve prescribed burning. Other examples of problematic species include common velvetgrass and colonial bentgrass (*Agrostis capillaris*), which also have been found to increase in Garry oak ecosystems following prescribed burning.

The reintroduction of a regular fire regime in Garry oak ecosystems is considered to be a key approach to restoring ecosystems and golden paintbrush populations. But, as suggested by the fire-response behavior of certain nonnative species, the recovery of golden paintbrush and its habitat entails more than simply burning away the invading plants. The complexity of recovery is increased by the fact that as amounts of woody material in golden paintbrush habitat rise as a result of fire suppression, the risk of especially large and devastating fires also grows.

Furthermore, given the history of fire suppression in the region and the alterations in golden paintbrush habitat brought about by the presence of nonnative species, it has become difficult to predict just how golden paintbrush may respond to fire. An accidental fire at Ebey's Landing during the growing season in 2002 wiped out half the golden paintbrush population at that site. In the years following the fire, the population continued to decline, effectively destabilizing it and opening the way for invasion by nonnative species. A second fire occurred at the same site in July 2007, caused by fireworks. Both fires not only killed plants but also charred the soil organic layer.

The fires at Ebey's Landing, while accidental, have lent support to a key idea that explains the relationship between fire and golden paintbrush, or rather the success of fire in aiding the species' regeneration: timing. Research has indicated that the seasonal timing of fire is critical to the success of prescribed burning for many species. In the case of golden paintbrush, the benefits of prescribed burning appear to be greatest when sites are burned in the fall, after plants have dispersed their seeds. By contrast, the growing-season fire of 2002 and the summer fire of 2007 at Ebey's Landing destroyed seed production for those years, thereby effectively crippling the population's growth in the following years. When burns are implemented in the fall, however, golden paintbrush populations tend to increase in size in the following year, the reason for their success being that the plants' seeds have dispersed. Prescribed burning has been used for golden paintbrush recovery at several sites, including at Rocky Prairie and at American Camp on San Juan Island, a site where the species was reintroduced in 2009. In both locations, burning has helped stabilize and encourage the growth of golden paintbrush populations.

Other forms of disturbance, namely mowing, have also been found to benefit golden paintbrush. Mowing of woody shrubs prevents the growth of shrubs to the point that they overshade golden paintbrush. But similar to fire, mowing is not a universal means of control for nonnative species, and of particular importance to its success, too, is timing. In the case of bull thistle, for example, mowing too early in the growing season encourages the plant to resprout. Likewise, while mowing can be used to limit the invasion of common velvetgrass, if mowed too early or too late in the year, the species will grow more vigorously. More effective, in some cases, is the outright removal of shrubs and other woody species, which produces immediate and lasting benefits for golden paintbrush, assuming that the shrubs and other woody plants do not return. At West Beach on Whidbey Island, the removal of rose shrubs in 2005 resulted in measurable population expansion of golden paintbrush during the subsequent growing season.

Hand weeding proved to be the ultimate solution to nonnative species invasion in a study of restoration strategies at Rocky Prairie. Experiments conducted there in the 1990s showed that the felling of Douglas fir, combined with the revegetation of bare areas with native forbs and Idaho fescue seedlings, had mostly positive effects toward restoring the native plant communities of golden paintbrush habitat.[14] But even under the carefully controlled conditions of the study, nonnative species challenged restoration efforts. Particularly problematic were bull thistle, colonial bentgrass, common velvetgrass, hairy cat's ear (*Hypochaeris radicata*), and wood groundsel (*Senecio sylvaticus*), all of which became rapidly established in the soil where Douglas fir had been harvested. The only way to eliminate the nonnative plants was to pull them out of the ground by hand.

Nevertheless, with knowledge of seed production and growth behavior for nonnative species, hand weeding can be highly effective. And it is that sort of effectiveness that is needed for the restoration of Garry oak ecosystems and the recovery of golden paintbrush and its native associates. Overall, prescribed burning appears to be the most promising single treatment across golden paintbrush sites. That is true particularly for golden paintbrush recovery strategies that rely on direct seeding into prepared habitat areas. But the combination of prescribed burning and other methods of invasive control, such as mowing and hand weeding, is even more powerful. Those methods can be further augmented by the use of herbicides and solarization. Herbicides can be especially effective for the control of nonnative plants, but they often pose significant risks to native species, particularly when applied in close proximity to native plants. In addition, the timing of herbicide application, similar to prescribed burning, mowing, and hand weeding, appears to be critical to its success in killing off nonnative plants. Solarization, which involves laying a sheet of plastic over an area, leverages the greenhouse effect to raise soil temperatures to lethal levels for seeds and rhizomes. Solarization eliminates the risk of spread of viable seeds from plants that have been removed by other means. It is, however, a small-scale treatment and thus not applicable to large areas of habitat.

Complicating the control of nonnative plants in golden paintbrush habitat are efforts to protect endangered or threatened butterfly species, particularly Taylor's checkerspot (*Euphydryas editha taylori*) and island marble (*Euchloe ausonides insulanus*). The case of Taylor's checkerspot is of particular relevance for golden paintbrush, which appears to be one of the butterfly's preferred host plants, serving as a site of egg laying and as a food source for larvae. But Taylor's checkerspot also feeds and lays eggs on the introduced species narrowleaf plantain, and in Oregon it feeds almost exclusively on the introduced plant, presumably because golden paintbrush was long ago eliminated from the region. As a major invading species in some areas of golden paintbrush habitat, narrowleaf plantain is an ideal target for removal. The impact of its removal, however, could be detrimental for both Taylor's checkerspot and golden paintbrush, which could be overwhelmed if the butterfly descended upon it in significant numbers.

In addition to addressing fire exclusion and nonnative species invasions, strategies are also being developed to help alleviate pressure from other threats facing golden paintbrush. For example, golden paintbrush is haunted by the prospect of habitat loss from development on privately owned sites. Several of those sites are zoned for residential development, thereby pitting golden paintbrush against a potential driver of economic growth. One way to overcome that challenge is for conservation specialists to work closely with landowners, making them a part of the conservation effort.

Such cooperative efforts are being forged in both Canada and the United States. One of the critical recovery criteria for golden paintbrush that was listed in the first US recovery plan for the species (published in 2000) concerned the establishment of at least fifteen populations on protected lands. Protected lands were defined as recovery sites that were owned or managed by a government entity or by a private conservation group concerned specifically with the recovery of golden paintbrush or that were designated under permanent conservation easements or agreements that brought landowners into the conservation picture. The Canadian recovery plan (published in 2006), in which golden paintbrush is part of a multispecies maritime meadow and Garry oak ecosystem recovery effort, placed significant emphasis on measures to encourage increased cooperation from landowners and habitat stewardship.[15] Public education and outreach are necessary components of maritime meadow recovery, particularly on Vancouver Island, where tension between urbanization and conservation has grown in recent decades.

In thinking about how best to approach the recovery of the Garry oak ecosystem within the confines of island boundaries in Canada, one can see where stewardship becomes a fundamental component. The Garry oak eco-system in Canada exists entirely on islands, which have only so much space to offer. Many of the people who live on those islands are living within areas of fragmented Garry oak ecosystem. They routinely interact with and affect the ecosystem's components, including its plants, animals, water, and soil. Therefore, no person living on southeastern Vancouver Island, the most densely populated portion of the ecosystem in Canada, is free from respon-sibility. All who own and use the lands must participate and cooperate in conservation and restoration strategies, if the ecosystem is to continue to function and to provide its human inhabitants with clean water, clean air, soil, and aesthetic appeal, as well as support the survival of endangered spe-cies. The need for that level of public involvement is made clear in the Canada–British Columbia Agreement on Species at Risk, which states: "Stewardship by land and water owners and users is fundamental to prevent-ing species from becoming at risk and in protecting and recovering species that are at risk."[16] Public outreach efforts proposed in the multispecies re-covery plan ranged from the distribution of information on maritime mead-ows and their threatened species, to the engagement of the public in projects aimed at identifying new species populations, to workshops and presenta-tions for local governments, landowners, schools, recreation clubs, and First Nations peoples.

In the United States, public participation has been encouraged for golden paintbrush through various avenues. In Oregon, for example, researchers with the Institute for Applied Ecology produced "Wanted" posters that de-picted golden paintbrush and offered a monetary reward for its discovery. The posters were put on display in locations visited by rural landowners on whose properties the species was most likely to be sighted. Articles show-ing and describing the species and its habitat were published in major regional newspapers in late spring to coincide with flowering and to facili-tate the plant's identification.

One of the most interesting and influential outreach programs that has involved the species has been the Sustainability in Prisons Project (SPP). SPP began with a partnership formed between Evergreen State College and the Washington State Department of Corrections and later expanded to Oregon, where the Institute for Applied Ecology and the Oregon State Correctional Facility formed a partnership. The SPP brings together scien-tists, conservation specialists, and incarcerated individuals to create a col-laborative and educational environment. The project has yielded intriguing results. In the case of golden paintbrush, research carried out in collabora-tion with inmates led to the realization that female Taylor's checkerspots prefer to lay their eggs on golden paintbrush or harsh paintbrush when compared with narrowleaf plantain.[17]

The finding that Taylor's checkerspot is inclined to use golden paintbrush as a host plant at sites in Washington could lead to a convergence of butterfly and native plant conservation within the region's Garry oak ecosystems. That convergence is already under way in Canada, thanks to the Canadian recovery plan, which simultaneously addresses the relationships between native maritime meadow plants and native butterflies.

Other threats to golden paintbrush that are being addressed in both countries include loss from trampling and overcollection. Measuring the actual impact of these activities is extremely difficult. The site most affected by trampling and overcollection, or at least the site where these activities have been monitored most closely, is Fort Casey State Park on Whidbey Island. There, trampling and flower picking have been chronic problems, difficult to prevent despite efforts to raise greater awareness of their impacts on rare plants such as golden paintbrush. Trampling is also a potential threat on Trial Island due to maintenance activities in areas around radio towers.

At various sites, golden paintbrush has further sustained damage from predation by herbivores, including livestock, rabbits, deer, voles, and insect larvae. The most significant damage is caused by small mammals, particularly rabbits and voles, whose presence in golden paintbrush habitat has been associated with increased thatch and shrub cover. Grazing and seed predation by small mammals are suspected of reducing the reproductive potential of golden paintbrush, with populations that are already stressed and in decline suffering disproportionately large reproductive setbacks. Herbivore predation has been addressed primarily through the caging of golden paintbrush plants to protect their leaves and seeds. The attraction of rabbits and voles to specific successional characteristics of golden paintbrush habitat suggests that methods to stem the invasion of woody species may also be effective in reducing the impacts of herbivory.

Two additional though less common threats include susceptibility to erosion and marine pollution, such as from a catastrophic spill. Erosion has contributed to the loss of golden paintbrush plants specifically at Ebey's Landing, where plants are situated on a steep hillside. In 2004 part of the hillside collapsed into the surf, bringing its overlying blanket of soil and plants down with it. At that particular site, there is little that conservationists can do for the species.

In both Canada and the United States, golden paintbrush recovery is focused first on the protection of extant populations. Conservation specialists aim to maintain at least current abundance for the existing populations.

They also hope to expand those populations or establish new ones by reintroducing plants to historical sites and introducing them to new sites within the species' historical range. Many of the aforementioned approaches to dealing with invading woody species, invading nonnative species, and herbivores are concerned primarily with the conservation of existing golden paintbrush populations and of newly established reintroduction or introduction sites, such as those at American Camp and in the Willamette Valley. Reintroduction, introduction, and augmentation are aimed primarily at rebuilding the species' numbers.

In the United States, each of these efforts also is directed specifically toward fulfilling the recovery criteria for the species that were laid out in the original recovery plan.[18] The first criterion indicated that delisting of golden paintbrush would be considered when at least twenty populations, with a five-year average population size of a thousand plants, had been established throughout the species' historical range. According to the five-year review for golden paintbrush, published in 2007, only two of the eleven extant populations of golden paintbrush were considered to be stable.[19] The two populations were the one at Rocky Prairie and the one on Canada's Trial Island. Although the population on private property in the San Juan Valley met the population size threshold, as of the last count there in 2003, its stability was unknown, since officials were later denied access to the land. The assumption is that another eighteen populations need to be established to fill the first recovery criterion for golden paintbrush.

Another four recovery criteria were listed in the 2000 plan as well, but only two of them were seen as being critical to the ultimate decision of whether or not to remove the threatened label from golden paintbrush in the United States. One of those critical criteria was the previously mentioned establishment of at least fifteen of the twenty required populations on protected lands. When progress on the species' recovery was reviewed in 2007, that criterion was met again only by the populations at Rocky Prairie and Trial Island.

The second of the critical criteria concerned the collection and storage of seeds representing genetic material from across the species' geographical range. That process, because it allows for the selection of genetically fit and diverse seeds and plants suitable for propagation, was seen as a fundamental force behind reintroduction. To allow for the greatest success for reintroduction efforts, researchers had to first secure genetic material and study it to gain knowledge of the genetic diversity and genetic structure of extant populations. One of the most insightful studies on the subject uncovered high levels of genetic diversity in golden paintbrush populations based on genetic analyses of leaf samples gathered from each of the eleven extant populations.[20] Mean genetic diversity was highest for populations with more than a thousand individuals, but even among small populations, factors that normally limit diversity, such as inbreeding, were of little significance. That inbreeding was found to be negligible is consistent with self-incompatibility in golden paintbrush. Self-incompatibility normally is a concern for conservation, but its potentially negative effects on golden paintbrush appear to be offset by the species' naturally high genetic diversity, despite its limited geographical occurrence.

Based on these findings, researchers have been able to make important progress toward recovery. For example, the development of production beds has led to a greater availability of golden paintbrush seeds, which in turn has enabled a shift away from the use of expensive nursery-grown plugs for outplanting. Although seedling survival rates remain low, direct seeding for golden paintbrush has proven successful when used in conjunction with various site preparations, such as prescribed burning followed by seeding with golden paintbrush and mixtures of native forbs and grasses. Other research has indicated that seed stratification to mimic natural wintering processes and supplementary watering in the first summer after seeding in recovery sites may result in increased seed germination rates and increased seedling survival, respectively. More and larger populations of golden paintbrush have been established at reintroduction sites in recent years. The speed at which that progress has occurred has led to the anticipation of a proposal from the US Fish and Wildlife Service to delist the species by 2018.[21]

The various strategies allow for the restoration of not only golden paintbrush but also prairie habitat. And that may be the most important factor that dictates the success of recovery efforts. Research on the effects of genetics and habitat quality in determining the performance of reintroduced golden paintbrush plants has shown that plants grown from wild seed sources survive best in sites that are similar ecologically to the sites from which the seeds were collected.[22] By contrast, genetic diversity, population size, and physical distance between seed source populations and reintroduction sites have little influence in determining the survival of directly seeded golden paintbrush plants. That those factors do not impinge on the success of golden paintbrush recovery is intriguing, particularly given their influence on the outcome of recovery efforts for many other at-risk plant species.

The multispecies approach developed for the recovery of maritime meadow plant species at risk in British Columbia's Garry oak ecosystems naturally lends itself to recapitulating ecological conditions at reintroduction

sites. The Canadian Garry oak ecosystem recovery plan includes the recovery of six other plant species in addition to golden paintbrush, each of
which, like golden paintbrush, has been affected by habitat loss. All also
continue to be threatened by ongoing urban development, recreational activities, nonnative species, loss of pollinators, and fire suppression.

Specific recovery goals have been developed for each species in the
Garry oak ecosystem recovery effort. For golden paintbrush, in addition to
maintaining current abundance for existing populations on Trial Island and
Alpha Islet, recovery goals included the establishment of seven new populations throughout the species' historic Canadian range. The range included
the Garry oak ecosystem along southeastern Vancouver Island, as well as
areas of that same ecosystem on certain islands in Haro Strait, the Strait of
Georgia, and the Strait of Juan de Fuca. Seven new populations would
make for a total of nine golden paintbrush populations in Canada. Once
those populations are in place, specific recovery goals for the species in
Canada will be fulfilled.

Ultimately, the success of recovery efforts for golden paintbrush in both
Canada and the United States will be defined by the stability of populations. Given the tremendous pressure of encroachment by nonnative species
and habitat loss and the unpredictable nature of some current populations,
achieving sufficiently large, stable populations of golden paintbrush poses
a significant challenge. In both countries, success will be further determined by the general well-being of golden paintbrush habitat. In Canada,
that well-being is described in terms of the recovery of the Garry oak ecosystem. In the United States, it is described in terms of habitat recovery. In
both cases, the attention given to the recovery of golden paintbrush and its
native associates could help restore one of the Pacific Northwest's lost habitats: its grasslands.

Better known for their mountainous landscapes, their inlets and fjords,
and their salmon and orcas, southwestern British Columbia and the coastal
northwestern United States are home to a diverse range of topographical,
geological, and biological features. But that diversity is not, perhaps, self-
evident, for there is a level of drama in the landscape that easily draws our
attention away from the less dramatic but equally compelling. The view,
for example, from the bridge over Deception Pass, which lies between Fi-
dalgo Island and Whidbey Island, captures that drama in a way that words
cannot. Best translated, the scene would be described simply as rugged. The
swirling waters below are eternally dark and churning, while the trees on
the land along the pass are the expected species—Douglas fir, hemlock,
spruce. The scene does not include meadows, much less meadows painted

with streaks of gold. But just beyond the pass, on Whidbey Island, unfolds a new landscape, one where meadows become possible, where we could imagine the existence of a species like golden paintbrush.

It is a fragile existence, but it is a real existence nonetheless. And it makes one wonder what Whidbey Island or Vancouver Island or the Willamette Valley looked like when golden paintbrush was at its height. In some ways, it must have been like the scene at Deception Pass, its beauty, its awe beyond words.

Hidden Value

The loss of plants in North America happens quietly, with many people seemingly unaware that the natural environment around them is degrading. But plants and healthy ecosystems underlie our survival. The significance of that relationship is, in many ways, encapsulated in the economic value of plants, a value that is little recognized in the context of human health and food security.

Fundamentally, all species are valuable, because biological diversity helps to ensure ecological stability. The more stable an ecosystem, the better it is able to respond to changing environmental conditions. The better it can cope with change, the more consistently it is able to carry out functions such as primary production, soil development, and water budgeting. Each of those functions is useful to us, and so they are often referred to as ecosystem services.

But where some see ecosystem services, others see land waiting to be cleared for development, and where some would go to great lengths to protect the species and biodiversity that make those services possible, others see little point in doing so. Human populations are still growing, after all, and so our demand on the environment will increase. In extreme cases, some people may even work to actively prevent species protection. The difference between those who support species conservation, those who do not support it, and those who are indifferent rests largely with what people perceive as valuable and why.

When it comes to saving and restoring rare and threatened species of plants, however, our personal perspectives are irrelevant. The fact of the matter is that we depend on plants of all kinds for our health and survival, which gives them value beyond most other things. That is a compelling statement, and one that many advocates of conservation readily turn to when expressing the need to protect plants. But it is also a statement about inherent value, and to believe it, we must be willing to place our trust in that kind of value. Furthermore, our world is not universally grounded in the inherent value of things. It is, rather, driven by the strength of economies. So simply saying that plants have great inherent value, even if we believe that they do, is not enough to ensure their protection. To give conservation a chance, the value of plants must be quantified in ways that are meaningful economically.

Just how economically important our native plants will be in absorbing the impact of predicted changes in factors such as climate, population growth, and land use remains unclear. Estimating the economic value of threatened or endangered plant species is extremely challenging, not least because it requires us to measure the value of individual ecosystem components. Breaking down ecosystems into their parts is problematic, since not all the components or their specific functions may be known. For endangered species, it is even more uncertain, since their reduced populations can limit their study and diminish their true ecological significance. Moreover, the services an ecosystem offers likely are greater than the sum of its parts.

Nevertheless, we must start somewhere. In the case of at-risk plant species in North America, we can begin by determining whether or not they have relatives that are of economic value. Many of the continent's rare and threatened plants do have economically valuable relatives, which suggests that the rare species are themselves important economic resources. In some cases, we already know that they are critical resources—without them, iconic ecosystems would fail, potential sources of new medicines would be lost, and possible solutions to potential food crises would disappear. Their value is not always readily apparent, however, and in fact, it often is cryptic.

A major problem for many rare and threatened species is that resources, from our perspective, are to be used. So while valuable as stores, they cannot provide their full economic benefit until put to use. But rare or threatened plants often are not sufficiently abundant in the wild to allow for their use directly. Harvesting, for example, whether for food, medicine, or timber, could decimate remaining populations. Their relatively low abundance may also bring into question the extent of their aesthetic impact on the landscape. Enlivening only a small patch of forest with color, as opposed to being a keystone species, for example, is unlikely to form a convincing argument for conservation, particularly when arguing for project funding.

The key, then, is to quantify in terms of dollars the benefits that at-risk plants bring to our society. To do so with rare and threatened plants, we need to look beyond inherent value, or what those who advocate for conservation often see as intrinsic value—the value of a species in and of itself. We also need to think bigger than aesthetic value and obvious extrinsic worth, such as that derived from harvesting and selling the harvested product. We need to look instead to what rare and threatened plant species contain within, their hidden extrinsic value. Their real worth to us, in other words, lies in their genes, their biochemical properties, their ecological roles, and their history. Each species highlighted in this book possesses hidden extrinsic value, and their hidden value is far greater than most people realize, owing not only to each species' ecological significance and what we can learn from its past and present existence but also to the fact that North America's native plants and rare plants are largely untapped genetic and biochemical resources.

To gain insight into the potential monetary value of those untapped resources, we can consider what is known about the value of similar resources in other species. A useful example is the economic impact of elucidating and studying the complete sequence of the human genome. The economic output generated between 1988 and 2010 by the effort to sequence the human genome has been estimated at $796 billion.[1] That amount, which included job creation and growth and activities in research and industry, was only slightly less than the value of the global pharmaceutical market in 2010. But while knowledge of the human genome sequence is of vital importance in medicine, it is of little consequence when it comes to helping humankind meet one of its most basic needs: food production. And although that knowledge can help in the development of new medicines, many of the compounds from which those medicines are derived were found through natural products drug discovery.

In the area of food production, an example of the importance of untapped genetic resources is provided by sorghum (*Sorghum*). Based on production acreage, sorghum is the fifth most important crop grown worldwide. It is a staple food for more than half a billion people in Africa, Asia, and Central America and is likely to become an increasingly important food source in those areas in the coming decades as demand for food grows. Sorghum is unique among crops because it is adapted to warm, dry growing conditions. But it is also susceptible to low yields, particularly when affected by persistent drought, diseases, or pests. Low yields are especially problematic in less developed countries.

One approach to boosting sorghum production is through genetic improvement, which can be accomplished through either breeding techniques or genetic engineering or a combination of the two. The crop's genetic potential has been revealed through whole-genome sequencing of cultivated sorghum (*Sorghum bicolor bicolor*), different races of cultivated sorghum, and different species of sorghum.[2] The genome sequence of cultivated sorghum appears to be derived from a complex history of domestication, with the species having undergone domestication at least twice. The genomic sequences of different cultivated sorghum races, as well as the sequence of the wild species *S. propinquum*, are marked by a high degree of genetic diversity. Those sequences also have been found to contain previously unknown genes and genetic variants, some of which may provide plants with important traits for cultivation, such as increased grain size and quality and drought or insect resistance. Thus, across the genomes of different races and species of sorghum, one finds a deep pool of genetic resources and potential for finding ways to improve sorghum productivity.

All the genes within a plant's genome work together to make that plant what it is. But when it comes to improving crop species, a single trait-modifying gene from a related wild species can carry significant monetary value. An example is a gene known as *rhg1* (Resistance to *H. glycines*) in soybean (*Glycine max*) plants that are resistant to soybean cyst nematode (SCN; *Heterodera glycines*). About 90 percent of commercial SCN-resistant soybean varieties sold in the central United States are derived from a line known as PI 88788, which expresses the *rhg1-b* allele, to which the line's SCN-resistant qualities have been attributed. Research has indicated that overexpression of three genes at the *rhg1* locus (chromosomal position), in the form of increased copies of the genes in soybean, enhances resistance to SCN.[3] Thus, the selection for soybean varieties that carry more copies of the genes at *rhg1* or the deliberate overexpression of the genes through genetic-engineering methods could improve SCN resistance. Because SCN causes an estimated annual loss in soybean crops amounting to $1 billion worldwide, and because copy number variants of *rhg1* have the potential to overcome that annual loss, the *rhg1* collective has been described by some as the "billion-dollar gene."

A potential billion-dollar gene in a soybean plant is remarkable, and it would be something to see the gene ultimately reach economic impact on that scale. But just knowing that it exists elevates the importance of rare species and varieties of plants, which likely contain many genes that could be used to improve the yield and survival of widely cultivated crops and other economically significant plant species or varieties. If identified and tested successfully, those genes could allow farmers globally to save billions more dollars on agricultural endeavors while increasing levels of food production at the same time. And while the idea of eating genetically modified foods is not popular with some people, we may find ourselves becoming increasingly reliant on them, particularly if we cannot find solutions to expand food production through means of increased crop yields with bred varieties or improvements in sustainable farming. For that reason alone we should be doing everything we can to save the genes of crop wild relatives, which also means protecting those species as they exist in the wild.

The existence of plant genes of such great value shares an intimate association with biodiversity. Biodiversity is underpinned by genetic diversity, which allows species to adapt to specific conditions and to express certain traits, such as naturally occurring SCN resistance in only certain soybean varieties. Plants that are native to Canada and the United States are, by their nature, biologically diverse, having competed for habitat and become adapted to specific conditions. That is most evident in a group of plants such as *Penstemon*, where a large number of native species exists, making for an amazing depth of biological and genetic diversity. But even in a small genus such as *Torreya*, where just two species exist in North America, each adapted to distinctly different parts of the continent, marked genetic diversity is possible.

The monetary value for most of our native plants, including crop wild relatives and plant sources of medicinal compounds in North America, is unknown. But there are several at-risk native species of plants whose economic potential is great, though awareness of their existence is lacking. Prominent among them are wild rice (*Zizania*) and the various species of coneflower (*Echinacea*). The first is a source of food, while the second comprises a rich reservoir of promising medicinal compounds.

Rice presents an interesting case for crop improvement. It is among the most widely consumed foods in the world, with more than three billion people relying on it as a staple food. Its cultivation can occur under a variety of conditions, such as in irrigated lowlands or in dry, rainfall-fed uplands. It also was one of the first of the crop cereals in which the possibility of genetic transformation was demonstrated. In 1989 a gene known as *hph* was successfully introduced and overexpressed in protoplasts of the cultivated rice species *Oryza sativa*.[4] The experiments were performed with the basic goal of determining whether or not rice could be genetically modified. Their success opened the door to a world of opportunity for the study of gene regulation, and the transgenic rice plants themselves became a model for the genetic engineering of crops.

The United States produces just under 2 percent of the world's cultivated rice supply. Its limited production is largely a result of the fact that the United States sits at the edge of climatic tolerance for rice—the two major species of cultivated rice, *O. sativa* and *O. glaberrima*, are native to tropical Asia and Africa, respectively. As a result, rice farms within the United States are concentrated primarily in four regions: the Arkansas Grand Prairie, the Mississippi Delta, the Gulf Coast of Texas and southwestern Louisiana, and California's Sacramento Valley. All four regions are notable for their relatively warm and generally moist environments, which now house extensive, irrigated rice fields.

But North America also is home to its own kind of rice, better known as wild rice (genus *Zizania*). Three wild rice species occur here: wild rice (*Z. aquatica*) and northern wild rice (*Z. palustris*), which are annuals, and Texas wild rice (*Z. texana*), which is perennial. All are aquatic or semiaquatic grasses. All also were historically important sources of food for native peoples, and they still are highly valued for that purpose. Northern wild rice is the primary type of wild rice that is grown for commercial use. It is cultivated or harvested from the wild in the United States and Canada and is cultivated in several other countries, including Australia, Austria, and Hungary.

Northern wild rice is native specifically to the Great Lakes region of Canada and the United States, but it is in decline in its native habitat due to the loss of its native wetland environment. Wild rice (*Z. aquatica*) is native to Quebec and Manitoba in Canada and to the Atlantic Coastal Plain and Appalachian regions of the United States. It, too, has become relatively rare across much of its range. Of the three species, however, only Texas wild rice is listed as endangered in the United States. It earned that designation in 1978 after the diversion of water from its aquatic habitat resulted in a drastic reduction in its populations. It now exists only in the upper San Marcos River in Texas. Similar to the three North American species, the fourth member of *Zizania*, Manchurian wild rice (*Z. latifolia*), is nearly extinct in its home range in China, though it is now also invasive in New Zealand.

As crop wild relatives of both cultivated northern wild rice and cultivated *Oryza* species, North America's wild rice plants are of great interest for expanding the rice gene pool. Each of the three *Zizania* species has something unique to offer in terms of its genetic properties. Texas wild rice, for example, possesses several highly desirable genetic traits, including high heterozygosity and allelic richness, suggesting a high degree of fitness (successful survival and reproduction).[5] Hence, genes from Texas wild rice could be used to improve upon and diversify cultivated wild rice. Likewise, the genome of northern wild rice has been explored as a so-called tertiary gene pool for *Oryza sativa*, being a somewhat distant but promising relative for genetic improvement.

In the 1990s, researchers attempted to expand the rice gene pool in part by creating hybrids between genetically distant species of the *Oryza* group, such as between *O. sativa* and *Porterasia coarctata* (*O. coarctata*), a species also native to parts of tropical Asia. Unfortunately, such distant pairings, while having the potential to significantly diversify the rice gene pool, generally resulted in the production of few, if any, fertile embryos. To overcome the limitations of breeding, researchers turned once again to genetic transformation, using techniques such as microprojectile bombardment, in which particles coated with the genomic DNA of northern wild rice were shot into cultured cells of *O. sativa* plants.[6] Some of the transformed cells successfully regenerated to give rise to *O. sativa* rice plants with certain morphological traits that resembled those of northern wild rice. Although the incorporation of specific genes using the bombardment technique was unpredictable, the experiments demonstrated that with a commonly employed transformation technique, it was possible to produce rice plants with potentially useful traits. In theory, those plants could then undergo selection for criteria such as increased viability or yield under different growing conditions. One such growing condition would be a cool climate, to which northern wild rice is specially adapted. The successful incorporation of genes conferring cold tolerance could facilitate the cultivation of *O. sativa* in northern regions of North America, thereby enabling greater levels of rice production on the continent.

By 2050 we may see a 100 to 110 percent increase in global crop demand.[7] The demand will be so substantial that even if we were to convert and work all remaining land suitable for farming, we still would not meet our food needs. As a result, instead of thinking in terms of land conversion for farming, those who are working to address the problem of global food production have become increasingly focused on intensification, or boosting crop production on existing agricultural lands. Compared with land conversion, intensification would minimize associated increases in greenhouse gas emissions and nitrogen use while also avoiding the need to clear additional land for agricultural needs, which would inevitably destroy yet more biodiversity.

Intensification of crop production is where the expansion of crop gene pools becomes especially important. And again, we can look to rice as a model. Most rice consumption occurs in Asia, where long-grain *indica*

(*O. sativa indica*) rice cultivars are predominant. The other major cultivars, the short-grain *japonica* (*O. sativa japonica*) cultivars, are grown mainly in East Asia.[8]

For all rice cultivars, grain yield is determined largely by spikelet number and fertility. Spikelets are the individual flowering units of grasses. In rice, spikelets develop into grains. They are attached to panicles, which are indeterminate, branching inflorescences found at the upper ends of the plant's stems. Typically, the more panicles, the more spikelets, and the more spikelets, the greater a plant's grain yield.

Traits linked to yield in rice have been traced to naturally occurring variations in multiple genes. In *japonica*, for example, certain variations in the genes APO1, DEP1, and SCM2 have been found to increase grain yield, whether through increasing the number of grains per panicle (all three genes) or altering panicle structure (APO1). In addition, some tropical *japonica* varieties grown in Indonesia display traits associated with high yield potential, including large panicles, large leaves, thick stems, and vigorous root systems. Those traits and the associated increase in yield potential in tropical *japonica* plants are controlled in part by a gene known as *spikelet number* (SPIKE). The introduction of SPIKE into *indica* plants, accomplished through recurrent backcrossing into *indica* cultivars, can increase the plants' grain yields by 13 to 36 percent.[9]

The introduction of the new high-yielding *indica* varieties into existing rice paddies could lead to significant improvements in rice production globally. And it would happen without the need for additional land conversion for rice cultivation and without genetic engineering. Whether those high-yielding varieties will make it to North America remains to be seen. But augmenting rice production on the continent comes down to more than specialized breeding and genetic tools. Food security in North America is also intimately associated with the conservation of our natural resources, wild rice being a supreme example.

Wild rice is a type of cereal native to North America. It is known for its distinct taste and nutritional value, being rich in fiber, protein, and certain minerals and vitamins. Wild rice has been cultivated since the early 1950s, when the first flooded paddy was created in Minnesota. Since then, Minnesota and California have become leading wild rice producers, responsible for about 99 percent of cultivated wild rice production in the United States. Idaho, Oregon, and Wisconsin make up the remainder of US production. Each year, those five states generate between about eight and twelve million pounds of wild rice ready for consumption. Some years see much higher totals, sometimes reaching more than twenty million pounds. Much of the wild rice produced in the United States is used in blends with white rice. Canada produces another one to two million pounds of wild rice annually. There, the major wild rice–producing province is Saskatchewan. Wild rice in Canada is harvested from natural sources such as lakes, rather than being cultivated.

Although the bulk of wild rice production acreage in the United States lies in Minnesota, which falls within the native range of northern wild rice, California paddies generate higher yields. The reason for that appears to be associated with factors that specifically affect field-grown wild rice in Minnesota. Indeed, while natural stands of wild rice in the state are able to withstand disease, their field-grown counterparts have been subjected to especially destructive epidemics of disease, such as brown spot and stem rot. Many of the leaf diseases that affect cultivated wild rice are associated with extended periods of exposure to warm, humid conditions. In addition, drought and flooding have caused partial or complete crop failures, and yields may be reduced by seed shatter, in which seeds are dispersed rather than retained once they reach maturity, and by lodging, in which plants

Figure 13. Northern wild rice (*Zizania palustris*) in its native habitat at Lake Alice, Lincoln County, Wisconsin. (Photo credit: Courtesy of the Great Lakes Indian Fish and Wildlife Commission)

with filled grain heads simply fall over, unable to support their own weight.

To overcome these problems, scientists have been investigating the genetic resources of wild rice plants that demonstrate resistance to seed shattering and disease and that produce panicle types associated with improved yield. Since the 1970s, multiple wild rice varieties have been developed, most by researchers at the University of Minnesota. Among those varieties are K2, M3, Voyager, and Purple Petrowski, all of which are considered to be medium to high yielding or high yielding. Purple Petrowski, one of the more recently developed high-yielding varieties, has the qualities of resistance to lodging, shattering, and fungal disease. The development of superior varieties for cultivation would not have been possible without the genetic resources available in wild plants.

Another problem that faces wild rice production is climate change. Climate change is likely to place more stress on wild rice plants as a result of changes in water levels, changes in temperature, increased disease and pest outbreaks, and more intense or more frequent storm events. Stress from these factors could expedite the loss of wild rice habitat, lead to decreased plant vigor in both natural stands and field-grown crops, and result in further declines in remaining natural stands. Climate change could also invite greater numbers of invasive plants into wild rice habitat. These impacts combined would likely cause drastic reductions in yield from natural stands and field-grown wild rice. For *Oryza* rice, researchers already know that yields decrease when nighttime temperatures are above normal: a 1-degree Celsius increase in nighttime temperature causes a 10 percent decline in grain yield.[10] That is not an insubstantial amount, and overcoming it will likely require the development of varieties of rice with greater tolerance to warm nighttime temperatures. Texas wild rice, which is naturally adapted to the hot summers and mild winters that characterize its habitat, could be a key genetic resource for heat tolerance, possibly for both northern wild rice and varieties of *Oryza*.

Improvement of wild rice in North America ultimately may be determined by demand. If there is little growth in demand for wild rice products, there will be little incentive to improve or expand wild rice cultivation. Reduced interest in the plant as a source of food could also affect conservation efforts, which have focused on the protection of wild rice shorelands and aquatic habitat. Conserving those areas is key to ensuring the continued existence of the native northern wild rice gene pool. It would be unfortunate if interest in pursuing wild rice as a natural North American food resource does not grow in the coming decades. Not only is it more nutritious than cultivated rice, but it is also native and intimately linked to the continent's heritage.

Wild rice has played a significant role in the lives of the Ojibwe (Chippewa) and the Menominee. The Ojibwe are one of the largest tribes of Native Americans and First Nations peoples in North America.[11] They occupy lands in the upper Midwestern United States and central Canada, a range shared with northern wild rice. Northern wild rice was a sacred plant in their cultures, at the center of tribal life. It was harvested annually in late summer, when tribe members would paddle into rice fields in birch-bark canoes. They took with them forked ricing poles, which they used to propel their canoes (and the design of which left the roots of rice plants unharmed), and a tool to knock rice kernels off the plants and into their boats. With the arrival of European Americans, the development of reservations, and the decline of northern wild rice, its role in the subsistence way of life was diminished. The eventual commercialization of wild rice also virtually eliminated its profitability at the small scales depended on by native peoples to supplement their incomes.

Despite all that has passed, modern native peoples continue to practice the rice-harvesting traditions of their ancestors. Collectively, they harvest a small amount each year, by some estimates only about 100,000 pounds. But of greater significance than the amount hauled in is the continuation of the tradition, which is symbolic—a reminder that we need not always look to distant lands for the things that sustain us. Wild rice is a gift to North America, one that has been long overlooked but the value of which, restricting ourselves to extrinsic value, must measure easily in the millions of dollars. Intrinsically, the value of wild rice exceeds measure.

The drug equivalent of the rare billion-dollar gene in the soybean is the blockbuster drug, a pharmaceutical agent that brings in $1 billion or more in annual revenue for the company that sells it. Such drugs are few and far between. Between them, however, are dozens of other medicines that are capable of maintaining or improving human health or even saving human lives. Some of the most important pharmaceuticals, based on how widely they are used and on how effective they are in treating a broad variety of what historically were menacing diseases, are the antibiotics. Before the discovery of antibiotics, of which penicillin, discovered in 1928, was the first, people with serious bacterial infections often died. In fact, had antibiotics not been discovered, you or I might not be here today. So antibiotics not only allowed people to live but in a way have also helped shape our recent lineages.

182 · *Hidden Value*

Penicillin was isolated from a type of mold, but other natural reservoirs, including plants, are also sources of important antibiotics. Many other groups of drugs have also played important roles in medicine, including pain relievers, anti-inflammatory drugs, anticancer drugs, vaccines, birth control pills, antipsychotic drugs, and antivirals. A number of these drugs have been discovered, synthesized, or manufactured in North America, but few have actually been discovered from the continent's native plants. The best-known example is taxol (paclitaxel), an anticancer drug derived from a compound discovered in the 1960s in the bark of the Pacific yew (*Taxus brevifolia*). Taxol is used in the treatment of cancers of the breast, lung, and ovary. At its peak revenue generation, prior to the arrival of generic agents on the market in the early 2000s, it brought in more than $1 billion annually for Bristol-Myers Squibb.

Although we have not seen the likes of a taxol discovered in a North American plant species lately, that does not mean that such a lucrative agent does not exist. Many species of plants native to North America were used medicinally by indigenous peoples, and multiple species, including black cohosh (*Actaea racemosa*, or *Cimicifuga racemosa*), bloodroot (*Sanguinaria canadensis*), common evening primrose (*Oenothera biennis*), and devil's club (*Oplopanax horridus*), are the sources of profitable herbal remedies or dietary supplements. While the safety and effectiveness of many of these products is unclear, and while the conditions that they are purported to cure or improve are often exaggerated, the emphasis on the role of native plant species as sources of traditional medicines and as popular herbal medicines has inspired their investigation as sources of compounds for drug development. Through the latter process, agents isolated from natural sources are subjected to rigorous protocols to prove their effectiveness in the treatment of specific conditions and minimize risks associated with their use.

Similar to the way in which the genes of crop wild relatives contribute to gene pools, the biochemical constituents of wild medicinal plant species contribute to the pool of natural medicinal resources. The importance of maintaining that pool, as it applies specifically to native plants in North America, is evident in the genus *Echinacea*. There are nine species of *Echinacea*, all native to eastern or central North America. The genus has a long history of use by Native Americans, who used various plant parts and different preparations of *Echinacea* to treat conditions ranging from cough and sore throat to toothache and eye infection to venereal disease. For much of the first part of the twentieth century, echinacea was included in the US National Formulary. In recent decades, herbal and dietary supplements made from *Echinacea* have become immensely popular among American consumers. In 2012 in the United States, consumers spent almost $5.9 million on *Echinacea* products.

Extracts of the aerial or underground parts of three species, eastern purple coneflower (*E. purpurea*), narrow-leaf coneflower (*E. angustifolia*), and pale purple coneflower (*E. pallida*), are commonly used in herbal echinacea preparations. The three species also are relatively abundant in their natural ranges and frequently are grown in home gardens. The other species of *Echinacea*, which generally do not feature in herbal products, vary in their abundance. Yellow coneflower (*E. paradoxa*), for example, is threatened in Arkansas but not listed as such federally. Tennessee coneflower (*E. tennesseensis*) was recently removed from the US list of endangered species, following its recovery in its native cedar glade habitat in central Tennessee. The only species of the group that retains a federal endangered listing is smooth coneflower (*E. laevigata*), which is found in sparse populations in Georgia, South Carolina, North Carolina, and Virginia.

All species of the *Echinacea* genus, despite their abundance or lack thereof in the wild, possess significant medicinal potential as true pharmaceutical agents. Much of that potential has long been purported by those who use *Echinacea* extracts to treat or prevent colds.[12] To find concrete answers about the medicinal properties of *Echinacea*, researchers have begun to look specifically at bioactive compounds, or substances that exert effects on cells. The findings of such studies have yielded intriguing results about the pharmaceutical promise of *Echinacea*.

Key chemical constituents that have been identified in *Echinacea* include compounds known as alkamides, caffeic acid derivatives (e.g., chicoric acid and echinacoside), polyacetylenes, and polysaccharides. These compounds are found in different concentrations in different parts of the plants, a factor that likely contributes to the marked variations observed in effectiveness between herbal products. Three of the different types of compounds, the alkamides, caffeic acid derivatives, and polysaccharides, are thought to be the major active constituents behind the apparent immunological activity of *Echinacea*, particularly the prevention and treatment of upper respiratory infections.

By 2005 a total of fourteen different alkamides had been identified from the roots of the three herbal species, eastern purple coneflower, narrow-leaf coneflower, and pale purple coneflower. Most of the alkamides were found to inhibit lipopolysaccharide-induced production of nitric oxide in mouse macrophages (cells of the immune system). Lipopolysaccharide is an endotoxin found in the cell wall in gram-negative bacteria that is capable of producing powerful inflammatory responses through the production of inflammatory mediators (such as nitric oxide). Importantly, *Echinacea* alkamides were found to inhibit lipopolysaccharide activity while having relatively insignificant effects on the viability of macrophages, indicating that the compounds may be able to exert anti-inflammatory effects without harming the body's immune cells. Alkamides isolated specifically from narrow-leaf coneflower are known to block the activity of the inflammatory mediators cyclooxygenase and 5-lipoxygenase based on the findings of in vitro experiments.[13] Cyclooxygenase, of which there are several varieties, is a target of nonsteroidal anti-inflammatory drugs, examples of which include aspirin, ibuprofen, and celecoxib.

In the 1980s caffeoyl compounds, which are radicals of caffeic acid, were celebrated as a new group of promising antiviral substances. Multiple caffeic acid derivatives have been isolated from *Echinacea*, though the major ones are chicoric acid and echinacoside. In 1988 those two compounds, isolated specifically from pale purple coneflower, as well as caffeic acid from eastern purple coneflower, were reported to exhibit dose-dependent antiviral activity against vesicular stomatitis virus in mouse cells.[14] At their most effective doses, however, all three compounds were more or less toxic to cells, impairing cell growth and DNA metabolism.

Chicoric acid, which has also been found in chicory (*Cichorium intybus*), common dandelion (*Taraxacum officinale*), and green coffee (*Coffea* spp.) beans, has since become of particular interest in the search for new treatments for HIV (human immunodeficiency virus), influenza, and other viral diseases.[15] Research has indicated, for example, that chicoric acid inhibits an HIV enzyme known as integrase. Integrase, which is produced only by retroviruses such as HIV, allows viral DNA to become a part of the host genome. When the host's DNA replicates, it makes new copies of the viral DNA, which then give rise to new virus particles. Integrase inhibitors, by blocking the incorporation of viral DNA into the host's DNA, can effectively block the spread of retroviruses. Such inhibitors can be used in combination with other antiviral drugs to provide a robust therapeutic attack against HIV. Chicoric acid also appears to interfere with the entry of HIV into host cells.[16]

The antiviral activity that has been described for chicoric acid parallels the compound's biological function in the plants that produce it. Among its described functions in plants, for example, are protection against insects and various viral, bacterial, and parasitic (e.g., nematode) infections. But plants produce chicoric acid in prescribed quantities and with specific potency, qualities that are not necessarily consistent with the production of effects in humans. Plants also sequester chicoric acid in different parts, such as in roots or inflorescences, depending on environmental factors. Thus the next stage of investigation of chicoric acid for drug development is the generation of synthetic analogs. Some of that work has already been carried out. For example, analogues have been developed through the substitution of catechol groups for caffeoyl groups in the chicoric acid structure.[17] When the synthetic compounds were tested for their ability to inhibit HIV integrase, they performed similarly to or better than their parent compound.

Whether synthetic or semisynthetic chicoric acid derivatives will be developed into marketed pharmaceuticals remains to be seen. But it has become increasingly clear that the bioactive compound as it exists in natural sources such as *Echinacea* is extremely complex, with its concentrations and location in plants varying according to factors such as environmental conditions and life stage. Those variations can occur within the same species or even the same individual plant. In sanguine purple coneflower (*E. sanguinea*), for example, plants initially concentrate the compound in their roots, only to later decrease levels in their roots and increase levels in their flower heads. There are also significant differences in chicoric acid levels in wild versus cultivated *Echinacea* plants. Those natural variations are behind the variations we see in herbal products sold over the counter. According to one study, chicoric acid levels in eastern purple coneflower extracts ranged from not detectable to 3,891 µg/ml and in capsules from not detectable to 34.6 mg/g.[18] The measured quantities did not necessarily match those declared on the products' labels.

The stark contrast between what an herbal product is said to contain and what it actually contains is one reason to be wary of over-the-counter herbal remedies. Overharvesting, which leaves species imperiled in their natural habitats, is another. Not all plant species of interest as herbal products are easily cultivated, and once brought under cultivation, their characteristics begin to change under the force of selection imposed by humans. Wild plants, on the other hand, maintain their special adaptations, enabling them to deal with the pests and infections that afflict them naturally. Their way of coping with those factors is often through the production of compounds like chicoric acid. Although not always a direct solution to treating human disease, those compounds represent a starting point for the development of semisynthetic or synthetic drugs that are perhaps more potent and safer than their parent compounds. And while it is a still imperfect and much more expensive and time-consuming process than collecting and preparing plant extracts for herbal remedies, modern drug development is at least subject to regulation, such that more often than not we know what we are getting. The investment in an actual treatment often is worth far more for our health than the hit-or-miss quality of herbal products.

That investment also can be worth immense sums of money for pharmaceutical companies. Some of that money may be returned to nature in the form of conservation programs. The stripping of bark from Pacific yews for the development of taxol sparked an intense debate over Pacific yew conservation and use in western North America. Bristol-Myers Squibb eventually signed cooperative agreements with the US Forest Service and the BLM in which the company provided funding for species inventories, conservation research, and the development of conservation guidelines for the yew. The agreements led to the Pacific Yew Act, introduced in 1992, which provided "for the sustainable harvest of the Pacific yew, or Pacific yew parts, in accordance with relevant land and resource management plans for the manufacture of taxol" and provided for the species' long-term conservation in the wild.[19] In the end, a semisynthetic form of the drug was developed, which resulted in far less demand for the collection of material from the species in the wild.

The Northern Plants

In the remaining stands of Fraser fir in the Southern Appalachians, we find a history of forest migration that reaches back twenty thousand years. We also find a cultural history, most evident in the species' role as a Christmas tree. The few surviving populations of running buffalo clover relate the story of the persecution of one of North America's most iconic mammals, the bison. Mead's milkweed paints a vivid picture of devastating loss on North America's prairies. In wild rice, we see the cultural value and economic potential of native plants.

The stories of these species remind us that North America's plants are living histories. They tell us about the continent's past, about its escape from the grip of glaciers and its transformation to a mosaic of temperate forests in the Southeast, prairie and semidesert in the West, and coniferous forests and tundra in the North. They tell us of the time of European settlement and what came afterward. They provide us with a unique perspective of the cultures and events that helped shape North America's modern human identity. Time runs through their stories.

In the North, in the boreal zone and the tundra, time is especially relevant.[1] Some species of plants found in those regions have existed there for tens of thousands of years, having taken refuge in ice-free regions during periods of glaciation. Among those species are conifers, the lineages of which are extraordinarily ancient. Conifers are notably abundant in the boreal, where they form some of the most extensive intact tracts of forest left on Earth.

But the peaceful and beautiful existence that those forests have long enjoyed is fading away. For many species of plants that inhabit the boreal and tundra biomes, species herein referred to collectively as "northern plants," time is running out.[2] Northern ecosystems are changing. Some of that change has come about as a direct result of human activity, such as mining and development. Most of it, however, has been associated with climate change, the reach of which is inescapable. The entire northern region of the world, including all its vast expanses of coniferous forests, its many bodies of water, and the permafrost that lies beneath, has been subjected to the impacts of a shifting climate. And in the North, climate change has been especially severe, more so than any place else on Earth. Climate modeling and investigations of the effects of climate change on coniferous boreal species have indicated that, relative to other parts of the world, warming over subpolar lands in Alaska, northwestern Canada, and northern Eurasia is disproportionately strong.[3]

Northern plants have already begun to respond to that warming. They have done so most obviously through profound shifts in growth, distribution, and susceptibility to factors such as fire, disease, pests, and drought. Arguably the most high profile of these responses have been the changes in distribution, which hint toward a northern escape for boreal species. Their migration is taking them into the tundra, where conditions are becoming increasingly warm and moist and thus suitable for boreal vegetation.[4] Over the course of the next century, as much as one-third to two-thirds of tundra vegetation could be replaced by boreal vegetation. As species of boreal plants move north, the regions they vacate could be overtaken by the types of ecosystems that currently lie to their south.

If these patterns of migration hold true in the coming years, then northern ecosystems will change not only in their climatic features and species compositions but also in their ability to carry out their currently vital roles in global climate regulation. The boreal biome alone stores between one-fifth and nearly one-third of the world's terrestrial carbon dioxide, trapping it in sinks such as living biomass, dead organic matter, the soil layer, and peatlands.[5] The biome's uptake of carbon is balanced with carbon release, with the major determinants of that balance being rates of plant growth, decomposition, permafrost formation, and the frequency and severity of fires. The ecosystems that currently lie to the south of the boreal accumulate carbon at a slower rate than boreal ecosystems. A decreased rate of carbon storage, combined with the loss of carbon from the boreal zone as its vegetation moves northward, could contribute to the conversion of the region from a carbon sink to a carbon source, possibly, in turn, accelerating climate change.

But there is much that is unpredictable about the northern environment. The response of the boreal's plants to climate change, for instance, is unlikely to be uniform. The fate of northern plant species further depends on the presence and severity of other threats, the effects of which could act in synergistic fashion with climate change.[6]

The threats, the declines, and the recovery efforts that exist for northern plants, along with the numerous possibilities for migration, extinction, and adaptation under warming climatic conditions, are immensely complex. And so, here, to break down that complexity, we will focus on just two endangered species of the North. One of them, Furbish lousewort (*Pedicularis furbishiae*), sits at the southeastern margin of the boreal in the transition zone between a southerly temperate hardwood ecosystem and a more northerly boreal ecosystem. The second, hairy braya (*Braya pilosa*), is found in the Far North, at the northern boundary of the tundra plains. Both species occur in places of remarkable geography and ecology, where the land is as diverse as the life it supports. In that diversity of life is to be found a great array of survival strategies, some of which are entirely unexpected.

Furbish lousewort is no stranger to the unexpected. It grows along the banks of the Saint John River, a waterway that marks the border between southeastern Canada and the northeastern United States. There it is found almost entirely on the south side of the river, which is shaded, whereas the northern bank, primarily in Canada, is in full sunlight for most of the day. It is restricted to a habitat that sits between the river's spring high-water level and the forest floor at the bank's upper edge, a band that measures roughly 6 feet in width in most locations. Potential sites capable of supporting that thin band of habitat are found along a 140-mile-long stretch of the river, beginning about a mile and a half above the Big Black River confluence in western Maine and terminating near the Aroostook River confluence in western New Brunswick. The geographic restriction is extreme, so much so that Furbish lousewort bears the distinction of being a narrow endemic species.

That is an exceedingly rare designation, but one not entirely unfamiliar to the New England–Acadian ecoregion, the name given to that pocket of North America, which is known particularly for its high degree of species richness. Throughout much of the range of Furbish lousewort, the temperate-to-boreal transition is evident. Where more southern-affiliated species such as red oak (*Quercus rubra*), silver maple (*Acer saccharinum*), and white ash (*Fraxinus americana*) begin to reach their northern limits, species such as balsam fir (*Abies balsamea*), red spruce (*Picea rubens*), and white spruce (*P. glauca*) become increasingly predominant. The result is a remarkable assemblage of broadleaf and coniferous plants, some of which, like Furbish lousewort, thrive specifically in the riparian habitat along the Saint John River.

A major reason why Furbish lousewort and certain other species of plants have taken such a liking to the river's riparian zone has to do largely with hydrology. On the surface, the relationship seems counterintuitive. The upper Saint John River has little in the way of headwater storage, where lakes or other natural upstream reservoirs are able to regulate the river's flow. The river also houses the longest free-flowing stretch of water in northeastern North America. The lack of flow regulation and the long run of free-flow mean that the Saint John River experiences substantial fluctuations in its levels throughout the year. The greatest of those variations occurs in the spring, when the river's banks are inundated with floodwaters and scoured by flowing ice.

Flooding and ice scouring erode bank habitat and cause entire portions of the bank to slump away, and thus they are natural forms of habitat disturbance. In the short term, they are devastating, stripping the river's banks of soil and vegetation. They clear out plants like Furbish lousewort, they eliminate shrubby thickets, and they prevent the establishment of trees. They also leave behind new substrates, causing the formation, for example, of vertical accretion deposits, in which only certain types of plants can grow. The changes are nothing short of dramatic, and yet they are absolutely necessary for perpetuating the existence of Furbish lousewort.

Indeed, although populations of Furbish lousewort plants may be demolished by major flooding and scouring events, the disturbance creates the species' habitat. The moist environment, the lack of woody species that risk overtopping and shading out seedlings, and the presence of specific substrates, such as well-drained glacial lacustrine deposits, gravelly glacial drift (usually near groundwater seeps), or vertical accretion deposits, are associated with the success of seedling establishment for Furbish lousewort. But above all else, disturbance enables succession. The most ideal conditions for the growth of Furbish lousewort form about three to ten years after disturbance, when the affected area has been repopulated by midsuccessional species. The inherently high level of diversity of those species seems to be especially critical to the survival of Furbish lousewort for reasons that are yet unclear. One possibility may be related to the fact that Furbish lousewort is hemiparasitic, perhaps obligately so at the stage of seedling development, though parasitic connections with the roots of other plants in its habitat have never been observed.

The plant species that associate with Furbish lousewort come in a variety of forms, as one might expect in a transition zone between ecosystems. Associates that are part of the region's boreal plant community include red spruce and white spruce, paper birch (*Betula papyrifera*), quaking aspen (*Populus tremuloides*), and the spreading shrubs green alder (*Alnus crispa*) and speckled alder (*A. incana* ssp. *rugosa*). Eastern white pine (*Pinus strobus*) is a common southern temperate associate. Additional shrub associates include bayberry willow (*Salix myricoides*), broadleaf meadowsweet (*Spiraea latifolia*), northern bush honeysuckle (*Diervilla lonicera*), red osier dogwood (*Cornus stolonifera*), and silky willow (*Salix sericea*). Bluejoint (*Calamagrostis canadensis*), eastern hemlockparsley (*Conioselinum chinense*), and Labrador Indian paintbrush (*Castilleja septentrionalis*) are examples of species found in the herbaceous layer in Furbish lousewort habitat.

Within that assemblage of species, Furbish lousewort is conspicuous. It emerges each May, its rosette of distinctively fern-like leaves coming to life after spring floodwaters have receded. Plants that are young and nonreproducing persist as small rosettes through the summer, while more mature, reproductive individuals, usually at least three years old, produce anywhere from one to more than a dozen dark, leafy, flowering stems, which typically appear in July. The stems, also known as flowering spikes or scapes, reach heights of 1.5 to 3 feet and change from green to a crimson shade of red as summer advances. They often are covered in short, sometimes conspicuously white hairs, as are the edges of the toothed and deeply lobed lanceolate leaves. On fully grown plants, the leaves are widely spaced and measure between 2 and 7 inches in length. They generally are positioned alternately on the stems, which also bear one or more tightly packed flower clusters. Each of the flower clusters contains about two dozen individual yellow flowers, which are remarkably similar in appearance to those of snapdragons.[7] They are, however, morphologically unique. Each flower is about one-half to three-quarters of an inch in length with a tubular corolla surrounded by five sepals, forming a five-lobed calyx. The corolla itself is without a noticeable beak on its upper lip (or galea), which together with the five-lobed calyx helps to distinguish Furbish lousewort from other northeastern species of *Pedicularis*, as well as snapdragons.

Flowering begins around the middle of July and lasts until late August, with the flowers opening in a staggered manner, such that only a small number present themselves at a time. Pollination is carried out primarily by the common wandering bumblebee (*Bombus vagans*). In the late 1970s, botanist L. Walter Macior noted that Furbish lousewort was one of the

Figure 14. Furbish lousewort (*Pedicularis furbishiae*). (Credit: Kara Rogers)

latest-blooming temperate species of *Pedicularis* to produce nectar. It was then also thought to be the only nectar-producing member of the genus to be pollinated solely by bumblebee workers.[8] In order to reach the nectar, bumblebee workers slip their proboscis between the galea and lower lip of the corolla. To accomplish that task, the bee holds onto the flower, flips itself upside-down, and swings backward toward the flower tube. As it sticks its head into the tube to drink the nectar, it effectively takes a bath in the flower's pollen. Wandering bumblebees are the plant's primary (or possibly only) pollinators, because they not only are common at the time of flowering but also are the only species whose proboscis is long enough to forage efficiently on the plant. Workers likely carry out the task because by the time the flowers are in bloom queens are engaged in brood rearing. The only other species of bumblebee that has been seen to remove nectar from Furbish lousewort is the now rare yellow-banded bumblebee (*B. terricola*), though it is not known to pollinate the flowers in the process of nectar feeding.

Pollination of Furbish lousewort can be successful regardless of whether the plants are self-pollinated or outcrossed with other individuals. When successful, pollination leads to the production of small fruits that yield tiny gray seeds following dehiscence. Although dehiscence occurs in late August or September, the split capsules usually remain attached to their scapes until the following spring, when the seeds fall into and drift away on the water or are scattered by the wind. Wind and water are thought to be the primary seed-dispersal agents for Furbish lousewort, but not all of its seeds are dispersed, as some simply fall to the ground beneath their parent plants.

There is no evidence to suggest that Furbish lousewort reproduces asexually in its river terrace habitat, which means that the effective production and dispersal of seeds are critical to its survival. In general, the species experiences relatively high levels of seed set. But each step along the way to securing the survival of those seeds and their development into seedlings is fraught with difficulty. More than one-tenth of maturing seed capsules may be lost to predation by larvae of geranium plume moths (*Amblyptilia pica*), and as much as one-quarter of a population's reproductive potential may be lost before or after seed set to herbivory by mammals. For seeds that make it through capsule set and predation and that are dispersed away from their parent plants, the difficulty lies in finding their way to suitable habitat.

But it takes just one Furbish lousewort seedling to originate a new population.[9] And with one seed able to give rise to an entire population, many new populations potentially can be established each summer. That potential is significant for several reasons, one being its link to the species' past.

Furbish lousewort likely dispersed into the Saint John River watershed some ten thousand to twelve thousand years ago. The dispersal event, in which seeds likely came to the river by way of either western North America or Asia (via the Bering land bridge), may have brought just one or possibly a few seeds to the river's banks. The establishment of populations from so few individuals would help to explain the species' low levels of genetic variation. The hub of *Pedicularis* diversity is in the eastern Himalayas, so it is probable that Furbish lousewort is a direct descendant of an Asian species.

The ability of a single seedling to give rise to a population and the species' consistently high seed set also suggest that Furbish lousewort has staked its survival in numbers. Its populations essentially are ephemeral, a fleeting character that arises from their susceptibility to extinction from stochastic, or chance, events, examples of which include disturbance from spring flooding and ice scouring. So the more seeds that Furbish lousewort plants produce, and the more efficient those seeds are at establishing new populations, the better the species' chance of surviving devastating disturbance events.

Each new population that forms following disturbance and dispersal can be thought of as representing a node, or genetic link, within the network of a larger population, or metapopulation. Species that survive as metapopulations tend to occur in areas where habitat is fragmented. The thin band of river terrace habitat occupied by Furbish lousewort is fragmented naturally by disturbance. As a result, rather than existing as one long continuous strip along the Saint John River, the habitat is divided into patches that undergo periodic turnover, with the loss of a patch at one site being counterbalanced by the creation of a new patch at another site. Where one patch is eliminated, a local population of Furbish lousewort may vanish entirely. But where a new patch becomes available, a new local population may emerge.

The continual establishment of downstream populations of Furbish lousewort implies the existence of upstream source populations. The Saint John River flows northward, from Maine to New Brunswick, which means that Maine, where more than 85 percent of Furbish lousewort populations are located, is the most likely home of source populations. To what degree local upstream populations in New Brunswick serve as sources for downstream populations there is unclear. The possibility for long-distance dispersal, via floating down the river over the course of several days, suggests that New Brunswick populations can be derived from source populations in Maine.

Source populations, similar to the local downstream populations they spawn, are susceptible to extinction from disturbance or, in the absence of disturbance, from ecological succession or other natural factors. All populations also are susceptible to loss from artificial disturbances that eliminate potential habitat. Shoreline development along the Saint John River in New Brunswick, for example, has resulted in the loss or alteration of ideal river terrace sites for Furbish lousewort seedlings. Between the towns of Grand Falls and Perth-Andover, activities such as road and trail construction and residential and commercial development have resulted in the loss of more than 40 percent of tree-covered parts of the river's shoreline. The loss of trees likely has been detrimental for Furbish lousewort, since it depends on moderate shade for establishment and survival. In addition, invasive species such as reed canarygrass (*Phalaris arundinacea*), sweetclover (*Melilotus albus*), and woodland angelica (*Angelica sylvestris*) now cover extensive areas of the tree-cleared shoreline, preventing the establishment of native plants.

The threat most frequently mentioned in the context of Furbish lousewort is the alteration of hydrology stemming from the construction of dams and their associated reservoirs. Within the species' Canadian range, dams are found on the Saint John River at Grand Falls and on the Tobique and Aroostook Rivers, which feed into the Saint John below Grand Falls. The loss of free-flowing water and the filling of reservoirs in those areas probably have translated into a loss of potential habitat and thus a decline for Furbish lousewort. It is likely, too, that the ice-scour regime has been greatly altered in impounded areas. Outside of that, however, little is known about the actual impact of dams and reservoirs on the species' survival, which comes as a bit of a surprise when one considers the way in which an endangered species listing came about for Furbish lousewort.

Historically, the range of Furbish lousewort probably was larger than the area now recognized. It may have stretched several miles farther south of the most southerly populations now known (near Perth-Andover) and might have included the lower portions of some of the river's tributaries, such as the Tobique and Aroostook. Evidence for that expanded range is based primarily on historical herbarium vouchers. Specimens that were collected in 1882 by J. E. Wetmore, who appears to have discovered the plant along the banks of streams that run into the upper Saint John, and in 1884 by Canadian botanist James Vroom, who gathered it in the vicinity of the Aroostook River, indicate, for example, that Furbish lousewort may have once inhabited the riparian zones of the river's tributaries. The legitimacy of the plant's occurrence in such areas, and particularly the Aroostook River, was supported by the collection of a specimen at the mouth of the Aroostook in 1901 by American botanist Joseph Richmond Churchill. Vroom's find and that in 1882 of his fellow countryman George Upham Hay, who identified the plant at a site in Perth-Andover, lend support to the idea that Furbish lousewort at one time occurred farther downstream in that region than it does today.

Historical records also give some indication of the areas where Furbish lousewort has experienced marked declines since the late nineteenth century. The portion of the Saint John River that lies between Fort Kent and Van Buren, for example, entertains only a few individual Furbish lousewort plants, though historically it is referred to on several occasions. That stretch seems significant particularly with regard to specimens found near Fort Kent, which is where the earliest reported collection of the species took place. The first specimen was found in the summer of 1880, when American botanist and artist Kate Furbish came across the plant on the river's bank in the vicinity of Fort Kent. At the time, Furbish was traveling through Aroostook County with the intention of gathering wildflowers, "hoping to add a few new species to the large number which I had already collected in various counties in the State."[10] When she saw the peculiar lousewort, she thought that she might have found a new species to add to her collection, and so she decided to send a specimen to Sereno Watson, curator of the Gray Herbarium at Harvard University. While Watson worked to identify the plant, Furbish recorded her description of it in 1881 in the *American Naturalist*, writing, "*Pedicularis* n. sp.? grew three feet high on the bank of the [Saint John] river where the water trickled down its sides."[11] The next year, Watson declared that the plant was in fact a species new to science, just as Furbish had suspected. He captured her role in the species' discovery by assigning it a Latin name in honor of her.

In 1946 another specimen was recovered from a site at Fort Kent, one worth remembering because it later came to mark what many believed was the last sighting of Furbish lousewort in the wild. By the 1940s the species had experienced declines in at least several parts of its range. At Grand Falls, for instance, occurrences of the plant that were known from specimens collected in the late nineteenth and early twentieth centuries had disappeared by 1943. Further declines were suspected along the segments of the river that lie several miles upstream and downstream of Grand Falls, owing to the 1931 construction of the hydroelectric dam there. By 1975, so far as could be determined, Furbish lousewort had disappeared throughout its range in both New Brunswick and Maine. In a report published that year, the Smithsonian Institution concluded that the species was probably extinct.[12] For a time, then, it seemed that Furbish lousewort was history, rather than a living history.

There was one person, though, who had not forgotten about Furbish lousewort, and it is thanks in large part to his keen eyes that the species is still here today. The year after the Smithsonian's report was released, University of Maine botanist Charles D. Richards was contracted by the US Army Corps of Engineers to carry out a survey of rare plants in the Saint John River watershed. The plan was to incorporate his findings into an environmental impact statement that the corps was developing to support a proposal for the construction of a hydroelectric facility on the river's upper portion in Maine. That endeavor, authorized by the US Congress in 1965, was known as the Dickey-Lincoln School Lakes Project.

It seems unlikely that anyone at the time could have predicted what happened next, though it probably unfolded in Richards's mind the moment he identified Furbish lousewort on the river's bank. He discovered not just one Furbish lousewort in the course of his survey work but some two hundred individual plants scattered among six populations. All were found growing along the stretch of the Saint John River in Allagash, Maine, that sits about 30 miles upstream of Fort Kent. His rediscovery of the species effectively broke the deadlock that had held up progress on the Dickey-Lincoln proposal for more than a decade. From its outset, the proposal had been controversial, primarily for economic reasons. But it also posed serious threats to northern Maine's environment, where it was planned to consume nearly 200 square miles of forest, wetland, and water areas.

While the economic faults ultimately were cited as the cause of the proposal's failure, Kate Furbish's little plant and Richards's rediscovery of it added significant leverage to the environmental argument against the project. In the year after the species' rediscovery, additional surveys of the bank habitat along the Saint John River turned up more plants, and by the year after that, 1978, a total of some 880 had been counted. But that was all, and roughly 40 percent of those plants were in sites that fell within the proposed impoundment area of the Dickey-Lincoln project. As a result, they were deemed to be highly vulnerable to extirpation. That risk factored into the decision by US officials in 1978 to add Furbish lousewort to the country's list of endangered species, giving it the distinction of becoming one of the first six plant species in the United States to receive such a designation. Its listing helped deliver the final blow to the proposed hydroelectric concept, though lobbying by supporters dragged the proposal along until the mid-1980s, when Congress finally deauthorized it. It turned out, too, that Furbish lousewort had taken refuge in its native habitat along the Saint John River in New Brunswick, where it was rediscovered by Canadian biologist George M. Stirrett and colleagues in 1977, subsequent to the work carried out by Richards. Surveys and population counts performed there likewise turned up only small numbers of plants. The species was added to Canada's endangered list in 1980.

Because populations of Furbish lousewort cycle between colonization and extinction, estimates of overall population size can fluctuate dramatically from one year to the next as well as over longer time frames. In 1980 an estimated 1,115 plants survived, two-thirds of which were in Maine and the remainder in New Brunswick. In 1989 nearly 6,900 flowering stems were counted in Maine, but a major ice-scouring event that occurred two years later reduced that number by more than half.[13] Since then, the highest count reported in Maine was in 2003, when a survey turned up just over 5,600 flowering stems. One of the lowest counts, down to about 2,300 flowering stems, came in 2005. In that period of time, the early 2000s, Furbish lousewort was identified at eighteen sites in Maine and three in New Brunswick. A report published in 2011 indicated that the Canadian population numbered close to 1,000 individuals.[14]

It is difficult to identify any single trend from these numbers, such as whether Furbish lousewort populations are increasing or decreasing. In New Brunswick, a recovery plan outlined in 2006 for the species included among its short-term objectives the establishment and implementation of protocols for data collection on population trends.[15] The ultimate recovery goal for the province was to increase population size and number of occurrences of Furbish lousewort. Specific ten-year objectives included the maintenance of existing populations between Grand Falls and Perth-Andover at 200 or more individuals each. For sites occupied near the Aroostook River junction and farther upstream, between Grand Falls and the US border, the goal was to maintain a minimum of 250 individuals in each population. Ten-year objectives also included the establishment of self-sustaining populations at additional sites along the river and the conservation of potential habitat. Since most sites inhabited by Furbish lousewort in New Brunswick occur on private lands, private landowner stewardship is an important part of the species' protection. At least fifteen stewardship agreements have been signed by private landowners. In addition, a small pocket of Furbish lousewort habitat located near Tilley, New Brunswick, gained protection within the George Stirrett Nature Preserve.

In the United States, the recovery plan, initially published in 1991, focused on downgrading the species' listing from endangered to threatened. That could occur once the species' population was maintained for a six-year period at a mean of seven thousand flowering stems, accompanied by the establishment of protected areas that encompassed at least half of the species' essential habitat, including occupied sites and potential habitat. Although habitat protection objectives had not been met in the river's lower reaches, the Nature Conservancy had conserved most of the species' occupied and unoccupied habitat in the portion of the Saint John River lying upstream of the Allagash confluence. By 2010 populations of Furbish lousewort were once again on the rise, though still short of fulfilling the seven thousand stems needed for downlisting.

Most descriptions of the threats facing Furbish lousewort have concentrated on the impacts of hydroelectric projects and the artificial alteration of its habitat along the Saint John River, ignoring entirely the possibility for impacts caused by climate change. Arguably, those impacts have crossed the line from possible to certain. An analysis carried out in the 1990s of long-term water flow and climate records for sites along the Saint John River uncovered a slight warming trend, with increases in mild winter days and winter rainfall. The rise in rainfall was linked to unexpected river flows during winter that were powerful enough to break up ice cover and to increased spring flooding and ice jams. Historically, winter rainfall and mild winter weather occurred only rarely in the region. Other research documented the occurrence of twelve ice-jam events along the Saint John River in the 1900s, with the majority having occurred in the latter decades of that century.[16]

Furbish lousewort populations may need as much as a decade to achieve optimal seed production, so disturbance intervals that span less than that amount of time have the potential to harm the species' survival. The increase in ice-jam frequency together with an increase in flooding frequency since the 1940s indicate that Furbish lousewort is subject to more frequent disturbance. One of the shortest population-reducing disturbance intervals recorded in the twentieth century lasted just six years, with an event recorded in 1984, followed by another in 1991. Although most major disturbance events for Furbish lousewort have been separated by periods lasting more than ten years, the potential for climate change to shorten those intervals, in the absence of an effective mitigation strategy, is increasing. Any single catastrophic event most likely would not devastate Furbish lousewort, since it exists as a metapopulation. But multiple events, occurring close together in time, could reduce the species to perilously low numbers. A part of history would be repeated, and we could be left once again to gamble on the long odds of rediscovery.

Conservation must address all the threats that face Furbish lousewort, and thus far it has succeeded in mitigating several major ones, including the construction of dams that would affect key parts of the species' habitat. But reducing the impact of climate change on the Saint John River remains

an elusive goal. Climate commands the greatest influence on the river's behavior, making climate change the dominant force in deciding the fate of
species endemic to the river's riparian zone. How much conservation can
succeed in helping the region's endemic species survive climate change is
unknown.

That uncertainty is even more salient, I think, for hairy braya, a perennial
species of the mustard family (Brassicaceae) that is endemic to Cape
Bathurst peninsula and the Baillie Islands, remote areas in the far north of
Canada's Northwest Territories. Hairy braya was first observed in 1826,
when Scottish explorer and naturalist Sir John Richardson discovered the
peninsula. Richardson noted a scent about the flowers that was comparable to that of lilac.[17] In 1848, while on an expedition to find Captain Sir
John Franklin, who three years earlier had disappeared in the Canadian
Arctic with his ships the *Erebus* and the *Terror*, Richardson again crossed
paths with hairy braya. The second time around, he collected a fruiting
specimen. Another such specimen was gathered two years later by William J. S. Pullen. Pullen, like Richardson, had been searching the region
for Franklin when he collected the plant at Cape Bathurst. The species
was not to be seen again until 2004, when American botanists James G.
Harris and Daniel L. Taylor found several hundred plants along the peninsula's coast. Hairy braya had escaped observation for more than a century
and a half, so much of the plant's natural history is unknown.

Hairy braya stands just under 5 inches at most when erect and sometimes is reduced to a prostrate condition, its stems hugging the soil. A single
plant may sport from one to more than two dozen hairy stems. Leaves,
when present, are similarly hairy. They typically are positioned at the stem
base. The upper portion of the stem supports dense clusters of white-petaled
flowers, which, following pollination by insects in midsummer, yield round
to egg-shaped hairy fruits that measure roughly two-tenths of an inch in
length. The light-brown seeds are small and do not appear to be well
adapted for long-distance dispersal. It appears, too, that the species depends
on cross-pollination for the generation of offspring. Once established, individual plants are long-lived, surviving perhaps as many as fifteen years.

Those biological and reproductive characteristics likely explain the species' confinement to the northern end of Cape Bathurst and the Baillie Islands. Unable to disperse any great distance, the species is essentially
trapped there. The peninsula and islands jut out into the Beaufort Sea,
which for much of the year is stilled by ice cover. Open water is seen only
intermittently through winter, when wind and water currents separate drifting

pack ice from landfast ice to form a flaw lead, and in the summer, when landfast ice melts and the flaw lead widens into the Cape Bathurst polynya in the Amundsen Gulf.

As part of the Arctic Ocean, the Beaufort Sea has been affected by changes in ice regimes in the Arctic. Since the 1970s, for example, the Arctic Ocean has experienced reductions in multiyear ice pack cover and shifts in ice motion and thickness. Summer sea ice extent in the Beaufort Sea is subject to interannual variability, but in agreement with other observations for the Arctic, there has been a general downward trend in ice extent over the sea since at least 1980.[18]

The decline of Arctic sea ice has no single, clear cause. Rather, the reasons for it seem to be tied to multiple factors, such as global trends in atmospheric warming, changes in atmospheric circulation, changes in cloud cover, shifts in patterns of ice motion, and warmer sea temperatures. The origin of all these factors may rest with climate change, but about that there is yet some uncertainty. There is consensus, though, that the amount of ice cover in the Arctic serves as an indicator of climate change. And for hairy braya, most everything about its survival comes down to climate change.

Hairy braya is an ancient species of the North. Its existence there dates at least to the last glacial period, which lasted from 110,000 to 11,700 years ago, a time that marked the final stage of the Pleistocene Epoch. Although generally presumed to have been blanketed in ice, parts of what are now Alaska and the northern areas of the Yukon and Northwest Territories actually were free of glaciers at that time. It was in an ice-free pocket there, the Bering Sea Pleistocene refugium, which encompassed the Arctic coast of modern-day Cape Bathurst, that hairy braya thrived, enabling its persistence into the current era. For that reason, the species is considered to be a glacial relict. The extent of its distribution across that refuge, however, and the size of its populations in that glacial era are unknown.

Today, the pocket of land across which hairy braya occurs is all of about 96 square miles in area. In 2012 thirteen subpopulations were described within that pocket, eleven of which were spread out specifically across the northwestern part of Cape Bathurst.[19] The other two were located in the north-central area of the Baillie Islands. Of those latter two subpopulations, one contained an estimated forty individuals and the other an estimated three hundred individuals. Most subpopulations on Cape Bathurst consisted of hundreds of plants, with one subpopulation exceeding an estimated ten thousand. Despite the promising numbers, however, several subpopulations were found to be at considerable risk of loss due primarily to accelerated coastal erosion.

Part of hairy braya's habitat is located along the coast, just short of where the flat, gradually rising plain of tundra ends, its edge abruptly slumping away into seacliffs. Other habitat sites sit farther inland, in uplands and along inlets and streams. All these sites, whether inland or along the coast, share in common similar soil characteristics, a major one being a relative lack of moisture. In fact, while subject to sea spray along the coast and flooding along inlets, and while surrounded by thaw lakes in poorly drained inland areas, hairy braya's patches of habitat are remarkably dry. They fall into the mesic-xeric range, xeric soils being very nearly desert-like. Looking across the species' habitat, with its barren areas and naked patches of calcium carbonate–rich sandy and silty clay loams, one gains a better sense of the mystery in hairy braya. Its habitat is like a discontinuous Arctic desert, where dry plots are scattered throughout an otherwise wet tundra plains ecoregion.

More than 45 percent of the soil surface in some areas of hairy braya habitat is bare, supporting only hairy braya and no other species of plants. Bare soils are thought to be essential to the braya's establishment. Many of its competitor species, in fact, are unable to cope with the kinds of processes that create and maintain those bare patches. Among these processes are disturbance by caribou, seasonal saturation with water, and surface erosion.

Perhaps the most significant of these factors is the seasonal saturation of depressions in the tundra surface. On Cape Bathurst and the Baillie Islands, the tundra surface is uneven. The irregularity is produced by seasonal uplift and settlement, in which the tundra surface is deformed by seasonal freezing and thawing in the active permafrost layer. Often, uplift and settlement occur in the basins of drained thaw lakes. Thaw lakes, also known as thermokarst lakes, are a tell-tale sign of thawing permafrost. Typically, the lakes form in depressions above ice wedges in the permafrost layer and are filled by permafrost meltwater. Many thermokarst lakes, after a period of growth, eventually drain, leaving behind depressed wetland basins. Such basins are common features on Cape Bathurst and the Baillie Islands, as well as throughout the rest of the tundra plains ecoregion, which stretches south and west to the Mackenzie River delta.

In hairy braya habitat, depressions in the tundra surface encourage the formation of pools of standing water at certain times of the year, such as during periods of snowmelt or rain. Few species of plants are able to survive the saturated conditions. Uplifts, or frost heaves, in the surface likewise keep soils barren by uprooting vegetation. On Cape Bathurst, trampling by caribou further rids the soil of vegetation. Caribou occupy the peninsula in the early part of summer, before hairy braya plants begin to emerge. The animals move into the area for calving and remain for a short while afterward. By the time they leave, the ground surface has been decimated, preventing the growth of some plants and enabling the growth of others, including hairy braya.

Among the few plants that are able to survive alongside hairy braya are Arctic willow (*Salix arctica*), entireleaf mountain-avens (*Dryas integrifolia*), and grasses such as Arctic bluegrass (*Poa arctica*) and Arctic wheatgrass (*Elymus violaceus*). Each of these species is remarkably tolerant of the dry soils. Arctic willow and entireleaf mountain-avens, which are dwarf shrubs, characteristically thrive in dry Arctic habitats. In far northern regions of the Arctic and at high elevations in more southern areas, their growth habits trend toward mat forming. Also inhabiting some of the same areas as hairy braya is its cousin smooth northern rockcress (*Braya glabella*), a species that likely evolved from the hybridization of two other brayas, hairy braya possibly being one of those parent species. In the areas where they overlap, hairy braya and smooth northern rockcress may hybridize to some degree. Another possible associate is Greenland northern rockcress (*B. thorild-wulffii*), a species once thought to be descended from hairy braya and limited to islands in the high Arctic. Genetic investigation suggests that hairy braya and Greenland northern rockcress may, in fact, be more closely related than expected.[20]

With characteristics of hairy braya habitat being dependent on seasonal freezing and thawing activity, climate is a dominating force in the species' survival. Permafrost thaw is central to the processes that create the plant's habitat, and it is a major factor underlying the species' greatest threat, coastal erosion. The thawing of permafrost facilitates denudation, or the wearing away of land. Along Arctic coastlines, permafrost thaw has combined with decreased sea ice extent and increased sea level, increased summer sea surface temperature, and increased storm surge and wave action to produce an accelerated rate of erosion. Along the northern edge of Cape Bathurst and on the coast of the Baillie Islands, waves have sliced deeper and deeper into the headlands each year. In the 1950s about 20 feet of land was lost annually to erosion along the Beaufort Sea coast. By 2011 that rate was estimated at 30 to 40 feet per year, and it was expected to continue to rise.[21] Some coastal subpopulations of hairy braya plants were positioned about 160 feet back from the shore in 2011.

In addition to speeding the loss of habitat at the coastline, climate-related changes in the Beaufort Sea have also placed inland habitat areas at risk. Rising sea levels and increased storm surge have increased the chance of

Figure 15. Eroding coastal habitat of hairy braya (*Braya pilosa*) showing ice-rich permafrost as the white area under the active layer. (Photo credit: James G. Harris)

flooding in inlets and streams, sites that house the majority of hairy braya subpopulations. In inland areas, these events carry with them an elevated risk of salinization, with receding seawaters leaving behind deposits of salt on the land. Coastal subpopulations likewise have been threatened by salinization due primarily to their increasing nearness to the water, where vegetation is bathed in sea spray in the summer. Hairy braya is not known to grow in high-salinity soils, and, in fact, some patches of its habitat are separated by areas of marked salinity.

Fortunately for hairy braya, the bulk of its subpopulations is located in areas that lie some distance from the coast. Still, about 15 percent of the species' total population of mature plants could be lost if at-risk coastal subpopulations succumb to erosion. So long as climate change continues, preventing the northern edge of the North American continent from falling into the ocean is an impossible task. And because the impacts of climate change cannot simply be removed from hairy braya's habitat, the effort to conserve the species will need to concentrate primarily on ensuring the protection of remaining subpopulations and exploring the development of a seed bank for the species. Some parts of the species' habitat are within areas recommended for management under the Tuktoyaktuk Community Conservation Plan, which provides guidance for conservation and resource management on Inuvialuit lands. Although not legally binding, incorporation into the plan of information on hairy braya could bring greater awareness in the region of the species' plight, potentially facilitating the development of a more robust conservation program. There also may be other subpopulations of hairy braya in unexplored areas of potential habitat on Cape Bathurst and the Baillie Islands. If that is the case, then the species may have a greater buffer against its threats than considered at present.

The state of hairy braya's at-risk coastal subpopulations is sobering. It is a very real instance of the difficulties that exist for the tundra's unique flora. Tundra plants and their habitats are being squeezed, with pressure coming from both the south and the north. Boreal species are descending from the north-facing slopes onto the tundra plains, while the northern edge of the tundra is disappearing into the ocean. At the same time, in inland areas underlain by permafrost, accelerated thawing of the upper permafrost layers is modifying habitat in ways that are irreversible. All these issues have come about as a result of human-induced climate change.

Perhaps the most worrisome aspect of climate change in the North and elsewhere is that its effects are being realized on a monumental and currently unconstrained scale. We have set in motion a complex series of events that cannot be reversed within our lifetimes or our children's or grandchildren's lifetimes. Research has indicated that even if carbon dioxide emissions stopped tomorrow, atmospheric temperatures would not decline for at least another thousand years. Likewise, the Intergovernmental Panel on Climate Change (IPCC) has warned that most aspects of climate change will persist over centuries and will be irreversible.[22] It would be too optimistic to expect that climatic impacts on species such as Furbish lousewort, hairy braya, and Fickeisen plains cactus could be lessened anytime soon.

Despite that unfortunate circumstance, for many species of plants that are under pressure from climate-related changes in their environments, at least some portion of their habitat is currently intact, and at least some subpopulations or individual plants remain a step or two removed from immediate threat of loss. Those relatively robust subpopulations and individuals are critical to species conservation, primarily because they provide resources for recovery. Those resources will not last long, but still, paramount among them is time. Healthy plants and subpopulations buy time for threatened species, allowing awareness to build, state or federal protection to be gained, and restoration and recovery measures to be identified and put into action. The amount of time available is fleeting, but for our most vulnerable species of plants, it could be enough to secure their continued existence.

Afterword

There is a melancholic edge to the stories of rare and threatened plant species. In some cases, there is a sense of despair that may compel us to resign species to their fates. After all, what can we do for a plant like hairy braya, which inhabits an environment being rapidly and permanently altered by climate-related changes?

Fortunately, species conservation efforts for plants are giving us answers to that question. The United States and Canada are home to about twenty-five thousand native plant species, about one-third of which are threatened. For many of those threatened species, an important step in their conservation is moving material into ex situ collections. Ex situ collections include seed banks, gene banks, pollen banks, materials stored in vitro, and living plants. Each of these resources can be used to aid conservation in at least one of several ways. Most significantly, they prepare us for the worst-case scenario, providing the materials necessary for the reintroduction and restoration of plant species that have disappeared from the wild. But they also can be used for research and education. In addition, with information on which species are protected ex situ and which are not, researchers are able to prioritize collection efforts, something they could not do very effectively prior to the emergence of collections data.

Ex situ plant collections in North America exist primarily at botanical gardens and organizations with a focus on plant science. Historically, data on these collections were isolated to the organizations that maintained the collections. There was, in other words, no collective knowledge of their resources. That changed in 2010, when data on the percentage of threatened plants in ex situ collections were consolidated through the North American Collections Assessment, an effort led by Botanic Gardens Conservation International (BGCI) US in partnership with the United States Botanic Garden and the Arnold Arboretum of Harvard University. The assessment tracked the number of accessible ex situ collections in Canada, Mexico, and the United States to facilitate planning and collaborative action for the conservation of North America's vulnerable and endangered plant species. The assessment revealed that 35 to 37 percent of threatened species in the United States and Canada were conserved in ex situ collections by 2010. A follow-up assessment three years later found that the figure had risen to 39 percent.[1]

That percentage is likely to continue to rise. The North American Collections Assessment was developed as a way of implementing specifically in North America the framework that was developed for the Global Strategy for Plant Conservation (GSPC).[2] It was designed with special focus on GSPC target 8, which concerned the creation by 2020 of accessible ex situ collections for three-quarters of threatened plant species, with at least one-fifth of those species being made available for recovery or restoration efforts.[3]

The GSPC, which became a cornerstone of plant conservation with its adoption by the Convention on Biological Diversity in 2002, owes its existence largely to botanic gardens, which played a key role in the strategy's development. The GSPC, in turn, has provided botanic gardens with a framework for plant conservation. In North America, in addition to their work in augmenting ex situ collections for threatened species, botanic gardens increasingly have embraced their role in community-based plant conservation efforts. That has been true particularly for urban areas, where interest in cultivating an appreciation for native plant species has grown, and for remaining natural areas of high plant diversity.

Botanic gardens and other organizations also have been working in other ways to engage the public. Examples include Project Budburst, which was initiated in 2007 by the Chicago Botanic Garden and National Ecological Observatory Network, Inc., and the Rare Plant Treasure Hunt, a citizen-science program initiated in 2010 by the California Native Plant Society. Project Budburst contributed to the development of the USA National Phenology Network (USA-NPN) Plant Phenology Program. In both projects, citizens, students, and researchers report their observations on plant phenology, such as the timing of first leaf or first flower. Their data are used by USA-NPN to monitor the effects of climate on plants, animals, and their habitats.

Project Budburst, the USA-NPN Plant Phenology Program, and the Rare Plant Treasure Hunt allow ordinary citizens to engage with plant science and plant conservation, much the same way that the *Thismia* hunt and the "Wanted" posters for golden paintbrush invited community involvement. They also provide avenues through which people can connect with the natural world. The personal relationship with nature that can develop from participation in a project that calls on citizens for their involvement is vitally important. When people search for rare species, record data on the timing of fruiting, or plant trees to restore habitat, they learn that they can make a difference for nature, contribute to their understanding of the natural world, and develop a meaningful connection to it. They come to know what it is that we stand to lose, and so they become invested in its protection.

The stories related in these pages also offer an opportunity to help people find meaning in North America's rare and threatened species of plants. In the end, it is our actions that will decide whether those species undergo extinction or thrive under protection. We can choose to learn, to raise awareness, and to participate in conservation, or we can choose to ignore the loss of plants and nature, at our peril. Fortunately, for many of the plants featured here, the balance is tipping increasingly toward awareness and protection, and from that we should draw inspiration. It is possible to bring threatened species back from the edge of a lost existence. Each time that we succeed in that feat, our future generations benefit, and that is something to celebrate.

Acknowledgments

The stories related in this book are very much the stories of the botanists and conservation scientists who have made it their life's work to know and protect North America's plants. I am especially grateful for the assistance given by the following individuals: Diana F. Tomback, John Frampton, Kevin M. Potter, John Cohen, Dan Kennedy, Philip Huber, John Bates, John E. Arnett, Jr., Joel T. Fry, Gerould Wilhelm, Linda Masters, Marlin Bowles, Cathy Pollack, Vivian Negrón-Ortiz, Kathy Robertson, Julie Crawford, Tom Kaye, Peter David, Mark McCollough, and James G. Harris.

Notes

Whitebark Pine

1. W. W. Macfarlane, J. A. Logan, and W. R. Kern, "Using the Landscape Assessment System (LAS) to Assess Mountain Pine Beetle–Caused Mortality of Whitebark Pine, Greater Yellowstone Ecosystem, 2009: Project Report," prepared for the Greater Yellowstone Coordinating Committee, Whitebark Pine Subcommittee, Jackson, Wyoming. Mortality ratings were based on visual examination of aerial photographs and application of a numeric mountain pine beetle–caused mortality (MPBM) rating system.

2. Ronald M. Lanner, *Made for Each Other: A Symbiosis of Birds and Pines* (New York: Oxford University Press, 1996), 22.

3. Only about 15 percent of whitebark seeds cached by Clark's nutcrackers are cached in sites ideal for whitebark germination. Most caching appears to occur at elevations too low to support whitebark. Gail Wells, "Clark's Nutcracker and Whitebark Pine: Can the Birds Help the Embattled High-Country Pine Survive?," *Science Findings* 130 (USDA Forest Service, Pacific Northwest Research Station, 2011).

4. The evolution of whitebark and its genetic diversity are reviewed in detail in Leo P. Bruederle et al., "Population Genetics and Evolutionary Implications," in *Whitebark Pine Communities: Ecology and Restoration*, ed. Diana F. Tomback, Stephen F. Arno, and Robert E. Keane (Washington, DC: Island Press, 2001), 137–53. The species' migration out of refugia, also pertinent to its genetic diversity and current distribution, is discussed in B. A. Richardson, S. J. Brunsfeld, and N. B. Klopfenstein, "DNA from Bird-Dispersed Seed and Wind-Disseminated Pollen Provides Insights into Postglacial Colonization and Population Genetic Structure of Whitebark Pine (*Pinus albicaulis*)," *Molecular Ecology* 11.2 (February 2002): 215–27.

5. Steve W. Taylor et al., "Forest, Climate and Mountain Pine Beetle Outbreak Dynamics in Western Canada," in *The Mountain Pine Beetle: A Synthesis of Biology, Management and Impacts in Lodgepole Pine*, ed. Les Safranyik and Bill Wilson (Victoria, BC: Pacific Forestry Centre, 2006), 71.

6. Jeffry B. Mitton and Scott M. Ferrenberg, "Mountain Pine Beetle Develops an Unprecedented Summer Generation in Response to Climate Warming," *American Naturalist* 179.5 (May 2012): E163–E171.

7. Background on declining snowpack in western North America is based on Joseph F. Casola et al., "Assessing the Impacts of Global Warming on Snowpack in the Washington Cascades," *Journal of Climate* 22 (2009): 2758–72; Philip W. Mote, "Climate-Driven Variability and Trends in Mountain Snowpack in Western North America," *Journal of Climate* 19 (2006): 6209–20; and N. Knowles, M. D. Dettinger, and D. R. Cayan, "Trends in Snowfall versus Rainfall in the Western United States," *Journal of Climate* 19 (2006): 4545–59.

8. Diana F. Tomback and Peter Achuff, "Blister Rust and Western Forest Biodiversity: Ecology, Values, and Outlook for White Pines," *Forest Pathology* 40 (2010): 186–225; and Diana F. Tomback et al., "The Magnificent High Elevation Five-Needle White Pines," in *The Future of High-Elevation, Five-Needle White Pines in Western North America*, ed. Robert E. Keane et al., *Proceedings of the High Five Symposium, 28–30 June 2010, Missoula, Montana*, Proceedings RMRS-P-63 (Fort Collins, CO: USDA Forest Service, Rocky Mountain Research Station, 2011), 2–28.

9. Robert E. Keane et al., "A Range-Wide Restoration Strategy for Whitebark Pine (*Pinus albicaulis*)," Gen. Tech. Rep. RMRS-GTR-279 (Fort Collins, CO: USDA Forest Service, Rocky Mountain Research Station, 2012).

10. Data from Matthew C. Hansen, Stephen V. Stehman, and Peter V. Potapov, "Quantification of Global Gross Forest Cover Loss," *Proceedings of the National Academy of Sciences* (26 April 2010), doi:10.1073/pnas.0912668107. Hansen, Stehman, and Potapov define forest cover in their analysis as "25% or greater canopy closure at the Landsat pixel scale (30-m x 30-m spatial resolution) for trees >5 m in height." According to the USDA's *Major Forest Insect and Disease Conditions in the United States: 2011*, there are more than 750 million acres of land classified as forest in the United States, with millions more acres of trees in urban settings. According to Hansen, Stehman, and Potapov, approximately 46,332 square miles of forest cover were lost in the United States, accounting for 6 percent of US forests and almost 12 percent of total gross forest cover loss for the year 2000 among countries with more than 386,102 square miles of forest cover, based on the Landsat scale.

11. Doubling in rate of tree mortality from Phillip J. van Mantgem et al., "Widespread Increase of Tree Mortality Rates in the Western United States," *Science* 323.23 (2009): 521–24. The researchers examined tree mortality in undisturbed areas of forest that were more than two hundred years old, thereby excluding the effects of "transient dynamics associated with stand development and succession."

Fraser Fir

1. For the geographic distribution of Fraser fir, see Kevin M. Potter et al., "Genetic Variation and Population Structure in Fraser Fir (*Abies fraseri*): A Microsatellite Assessment of Young Trees," *Canadian Journal of Forest Research* 38 (2008): 2128–37.

2. Despite its common name, Douglas fir (*Pseudotsuga menziesii*) is taxonomically distinct from species of true fir (genus *Abies*). The distinction is sometimes recognized with a hyphen in the common name (Douglas-fir).

3. The IUCN (International Union for Conservation of Nature and Natural Resources) Red List of Threatened Species is a comprehensive resource with information on the conservation status of plant and animal species worldwide.

4. The different scenarios describing the fate of the red spruce–Fraser fir ecosystem are examined in detail in P. T. Moore, H. Van Miegroet, and N. S. Nicholas, "Examination of Forest Recovery Scenarios in a Southern Appalachian *Picea-Abies* Forest," *Forestry* 81.2 (2008): 183–94.

5. Kevin M. Potter et al., "Evolutionary History of Two Endemic Appalachian Conifers Revealed Using Microsatellite Markers," *Conservation Genetics* 11 (2010): 1499–513.

6. The ancient climate probably was slightly warmer than the modern tundra, as suggested by certain paleo features of deciduous forests embedded within the ancient boreal. Changing patterns in North America's climate and vegetation from the Quaternary onward is discussed in Paul A. Delcourt and Hazel R. Delcourt, "Paleoclimates, Paleovegetation, and Paleofloras of North America North of Mexico during the Late Quaternary," in *Flora of North America North of Mexico*, ed. Flora of North America Editorial Committee (New York: Oxford University Press, 1993), 1:71–94.

7. Figures for boundaries between biological communities from Paul A. Delcourt and Hazel R. Delcourt, "Paleoecological Insights on Conservation of Biodiversity: A Focus on Species, Ecosystems, and Landscapes," *Ecological Applications* 8.4 (November 1998): 921–34.

8. Estimates of tree migration rates from Jason S. McLachlan, James S. Clark, and Paul S. Manos, "Molecular Indicators of Tree Migration Capacity under Rapid Climate Change," *Ecology* 86.8 (2005): 2088–98. Boreal migration predictions from J. R. Malcolm et al., "Estimated Migration Rates under Scenarios of Global Climate Change," *Journal of Biogeography* 29 (2002): 835–49. Rise in global surface temperatures from IPCC, *Climate Change 2007: Impacts, Adaptation and Vulnerability: Contribution of Working Group II to the Fourth Assessment Report of the Intergovernmental Panel on Climate Change* (Cambridge: Cambridge University Press, 2007).

Three Plants and Their Animals

1. Plantings on strip mines in some parts of the United States and on the prairie Sand Hills of Nebraska have expanded the range of jack pine, albeit not naturally.

2. Information on jack pine associates at Voyageurs National Park from Vilis Kurmis, Sara L. Webb, and Lawrence C. Merriam, Jr., "Plant Communities of Voyageurs National Park, Minnesota, U.S.A.," *Canadian Journal of Botany* 64 (1986): 531–40.

3. In some parts of its range, such as eastern Wisconsin, jack pine produces partially serotinous or nonserotinous cones, which may explain why jack pine has been able to regenerate without fire in recent times in some areas. Fire history in the Great Lakes states from Benjamin A. Sands and Marc D. Abrams, "A 183-Year History of Fire and Recent Fire Suppression Impacts in Select Pine and Oak Forest Stands of the Menominee Indian Reservation, Wisconsin," *American Midland Naturalist* 166 (2011): 325–38. Fire frequency near Mack Lake from A. J. Simard and R. W. Blank, "Fire History of a Michigan Jack Pine Forest," *Michigan Academician* 15 (1982): 59–71.

4. For jack pine management for Kirtland's warbler, see Philip W. Huber, Jerry A. Weinrich, and Elaine S. Carlson, "Strategy for Kirtland's Warbler Habitat Management," Michigan Department of Natural Resources, USDA Forest Service, and USDI Fish and Wildlife Service, October 5, 2001.

5. Data on jack pine in Wisconsin from Wisconsin Department of Natural Resources 2012 annual species summary, dnr.wi.gov/topic/ForestBusinesses/documents /JackPineReport.pdf.

6. The potential threat of mountain pine beetle to jack pine in Alberta is discussed in Catherine I. Cullingham et al., "Mountain Pine Beetle Host-Range Expansion Threatens the Boreal Forest," *Molecular Ecology* 20.10 (May 2011): 2157–71.

7. Physical features of *Trifolium stoloniferum* from US Fish and Wildlife Service, "Running Buffalo Clover (*Trifolium stoloniferum*) Recovery Plan: First Revision," US Fish and Wildlife Service, Fort Snelling, Minnesota, 2007. Some variability exists for the precise size of the flowering head, the flowering date, and stem height, all factors likely influenced by geography and variations in seasonal and annual climate.

8. A closely related species is buffalo clover (*Trifolium reflexum*), which, unlike running buffalo clover, is an annual, does not grow runners, and produces pink to purple-red flowers. It is also exceedingly rare. Its historical native range covered much of the eastern United States. Several states list it as endangered, and in two, Maryland and Pennsylvania, it is extirpated.

9. John M. Bates, "Frugivory on *Bursera microphylla* (Burseraceae) by Wintering Gray Vireos (*Vireo vicinior*, Vireonidae) in the Coastal Deserts of Sonora, Mexico," *Southwestern Naturalist* 37.3 (September 1992): 252–58. Bates's thesis, "Winter Ecology of the Gray Vireo *vicinior* in Sonora, Mexico" (1987), hinted at a possible mutualistic relationship between the gray vireo and the elephant tree.

10. Historical references to the specific location of Gray's elephant tree are hard to come by, but the region of the specimen is noted in *Proceedings of the California Academy of Sciences*, 4th ser., 18 (1929): 471.

11. Bioactive elephant tree extracts were described by E. Bianchi, M. E. Caldwell, and J. R. Cole, "Antitumor Agents from *Bursera microphylla* (Burseraceae) I. Isolation and Characterization of Deoxypodophyllotoxin," *Journal of Pharmaceutical Sciences* 57.4 (April 1968): 696–97.

Lost in the Wild

1. John Bartram eventually owned three farms in Kingsessing Township. In 1752 he gave his son James a portion of one of those farms, which lay to the north of the botanic garden. In 1777, after Bartram's death, James inherited the remainder of that farm. John, Jr., took over the main farm and the botanic garden in 1771 and later inherited the botanic garden, house, and surrounding farm. He continued to own and operate the farm but usually leased it in sections to tenants (Joel Fry, email communication to the author, 26 October 2013).

2. The events of the Bartrams' journey are retold in engaging fashion by Charles Jenkins, "The Historical Background of Franklin's Tree," *Pennsylvania Magazine of History and Biography* 57.3 (1933): 193–208.

3. John Bartram's diary entry describing his encounter with the Franklin tree (and another species he observed in the area) notes simply "severall very curious shrubs" (Nancy E. Hoffmann and John C. Van Horne, eds., *America's Curious Botanist: A Tercentennial Reappraisal of John Bartram, 1699–1777* [Philadelphia: American Philosophical Society, 2004], 197).

4. William Bartram, *Travels through North and South Carolina, Georgia, East and West Florida, the Cherokee Country, the Extensive Territories of the Muscogulges, or Creek Confederacy, and the Country of the Chactaws* (Philadelphia: Printed by James and Johnson, 1791), 467, accessed 6 October 2013, http://docsouth.unc.edu/nc/bartram /bartram.html. William Bartram visited the Altamaha River multiple times, but many of the details of his journeys through the region remain obscure. *Travels* is a heavily edited work and often combines several time periods into a single narrative (Fry, email).

5. *Gordonia franklini, Michauxia sessilis, Franklinia americana,* and *G. altamaha* were some of the names assigned to the Franklin tree in the eighteenth and nineteenth centuries. Lamarck published his version of the species' name in the second volume of his eight-volume work, *Encyclopédie méthodique: Botanique* (Paris: Panckoucke; Liège: Plomteux, 1783–1808), 770, accessed 8 October 2013, http://dx.doi.org/10.5962 /bhl.title.824.

6. The divergence time of *Franklinia* and *Schima* and population exchanges among Theaceae tribes were estimated by Mi-Mi Li et al., "Phylogenetics and Biogeography of Theaceae Based on Sequences of Plastid Genes," *Journal of Systematics and Evolution* 51.4 (July 2013): 396–404.

7. Franklin tree habitat from Francis Harper and Arthur N. Leeds, "A Supplementary Chapter on *Franklinia alatamaha,*" *Bartonia* 19 (1937): 1–13.

8. In general, Theaceae have flower parts in fives. The seedpods of *Franklinia* almost always have five main parts, but because the pods open at the top and bottom, ten cells can be counted.

9. Some botanists consider *Thismia* to constitute its own family, Thismiaceae. Those that continue to place *Thismia* in Burmanniaceae generally consider the genus as part of tribe Thismieae.

10. Ferdinand von Mueller, "Notes on a New Tasmanian Plant of the Order Burmanniaceae," *Papers and Proceedings of the Royal Society of Tasmania* (1890): 232–35.

11. For differences in mitreform inner perianth structure in *Thismia,* see K. R. Thiele and P. Jordan, "*Thismia clavarioides* (Thismiaceae), a New Species of Fairy Lantern from New South Wales," *Telopea* 9.4 (2002): 765–71.

12. Norma E. Pfeiffer, "The Sporangia of *Thismia americana,*" *Botanical Gazette* 66.4 (October 1918): 354–63.

13. Burmanniaceae reproductive strategies are described in P. J. M. Mass et al., "Burmanniaceae [Monograph 42]," *Flora Neotropica* 40/42, Saprophytes Pro Parte (17 April 1986): 1–189.

14. "Tropics Take Root Here: Scientists Hunt Rare Plant," *Chicago Sun-Times,* 22 January 1952.

15. A. A. Reznicek, "The Disjunct Coastal Plain Flora in the Great Lakes Region," *Biological Conservation* 68 (1994): 203–15.

16. Details on the *Thismia* hunt from personal communication between the author and Linda Masters (May 2014) and Gerould Wilhelm (October 2013). The 2011 hunt was organized by Masters and Rebecca Schillo Collings.

17. Martin Bowles et al., "Results of a Systematic Search for *Thismia americana* Pfeiffer in Illinois," report prepared for the Illinois Department of Conservation and the US Fish and Wildlife Service, 1994.

Mead's Milkweed

1. Data on remaining percentages of prairies from "Regional Trends of Biological Resources—Grasslands: Prairie Past and Present," USGS, Northern Prairie Wildlife Research Center, accessed 11 October 2013, www.npwrc.usgs.gov/resource /habitat/grlands/pastpres.htm.

2. Robert F. Betz, "Ecology of Mead's Milkweed (*Asclepias meadii* Torrey)," in *Proceedings of the Eleventh North American Prairie Conference* (1989): 187–91. According to Betz, letters exchanged between Torrey and Mead and specimens maintained at the Chicago Natural History Museum point toward the species having been renamed between 1846 and 1848. Some consider its formal renaming, however, to have occurred with Gray's publication almost a decade later. Asa Gray, *Manual of the Botany of the Northern United States*, 2nd ed. (New York: G. P. Putnam, 1856).

3. Betz, "Ecology of Mead's Milkweed."

4. George A. T. Hise, *One Hundred Years of History: Commemorating a Century of Progress in the West Liberty Community* (West Liberty, Iowa, 1938); and James Krohe, Jr., "The Breaking of the Prairie," *Illinois Issues*, October 1981, 19–24.

5. Depth of topsoil on virgin prairies from Robert F. Betz and Marion H. Cole, "The Peacock Prairie—a Study of a Virgin Illinois Mesic Black-Soil Prairie Forty Years after Initial Study," *Transactions of the Illinois Academy of Science* 62 (1969): 44–53. The extent of erosion on the Illinois prairie gained attention in the late 1980s with Robert F. Betz and Herbert F. Lamp's investigation of A horizon (or topsoil) depth in the state's virgin cemetery prairies ("Species Composition of Old Settler Silt-Loam Prairies," *Proceedings of the Eleventh North American Prairie Conference* [1989]: 33–39).

6. US Fish and Wildlife Service, with Marlin Bowles, ed., "Mead's Milkweed (*Asclepias meadii*) Recovery Plan," US Fish and Wildlife Service, Fort Snelling, Minnesota, 2003. See also Marlin Bowles, Jenny McBride, and Timothy Bell, "Restoration of the Federally Threatened Mead's Milkweed (*Asclepias meadii*)," *Ecological Restoration* 19.4 (2001): 235–41.

7. The Osage Plains also extend into central Oklahoma and north-central Texas, but those areas lie beyond the natural range of Mead's milkweed.

8. US Fish and Wildlife Service and Bowles, "Mead's Milkweed."

9. Diane L. Tecic et al., "Genetic Variability in the Federal Threatened Mead's Milkweed, *Asclepias meadii* Torrey (Asclepiadacea) as Determined by the Allozyme Electrophoresis," *Annals of the Missouri Botanical Garden* 85 (1998): 97–109.

10. Douglas A. Hayworth et al., "Clonal Population Structure of the Federal Threatened Mead's Milkweed, as Determined by RAPD Analysis, and Its Conservation Implications," in *Proceedings of the Seventeenth North American Prairie Conference: Seeds for the Future, Roots of the Past*, ed. N. Bernstein and L. J. Ostrander (North Iowa Area Community College, Mason City, Iowa, 2001), 182–90.

11. Douglass H. Morse, "The Turnover of Milkweed Pollinia on Bumble Bees, and Implications for Outcrossing," *Oecologia* 53.2 (1982): 187–96.

12. US Fish and Wildlife Service and Bowles, "Mead's Milkweed."

13. US Fish and Wildlife Service, "Mead's Milkweed (*Asclepias meadii*) 5-Year Review: Summary and Evaluation," US Fish and Wildlife Service, Chicago, Illinois, Field Office, Barrington, Illinois, 2012.

14. Issues surrounding the reintroduction of long-lived plants from Leonie Monks et al., "Determining Success Criteria for Reintroductions of Threatened Long-Lived Plants," in *Managing Eden: Plant Reintroduction's Promises, Perils, and Uses in a Changing Climate*, ed. Joyce Maschinski and Kristin E. Haskins (Washington, D.C.: Island Press, 2012), 189–208.

15. The new sites and the state of previously identified sites in Kansas for Mead's milkweed were reported by Kelly Kindscher et al., "A Natural Areas Inventory of Anderson and Linn Counties in Kansas," Open-File Report No. 158 (Kansas Biological Survey, Lawrence, 2009), 18.

Florida Torreya

1. Estimated five hundred to six hundred Florida torreya trees remaining in the wild from T. Spector, R. Determann, and M. Gardner, "*Torreya taxifolia*," in *IUCN 2014* (IUCN Red List of Threatened Species, version 2014.1, http://www.iucnredlist .org/details/30968/0). Some estimates have given a wider range, between five hundred and a thousand. Number of wild trees able to produce male or female cones from US Fish and Wildlife Service, "*Torreya taxifolia*; Florida Torreya 5-Year Review: Summary and Evaluation" (US Fish and Wildlife Service, Southeast Region, Panama City Field Office, Panama City, Florida, 2010).

2. Six centers of endemism in the Southeast, one of which is the Apalachicola region, from James C. Estill and Mitchell B. Cruzan, "Phytogeography of Rare Plant Species Endemic to the Southeastern United States," *Castanea* 66.1–2 (March/June 2001): 3–23.

3. Steepheads are characterized by the presence of a seep or spring at the base, which causes constant erosion.

4. US Fish and Wildlife Service, "Florida Torreya (*Torreya taxifolia*) Recovery Plan" (US Fish and Wildlife Service, Atlanta, Georgia, 1986).

5. Climate data for the Apalachicola region from "1981–2010 Normals, Apalachicola" (Florida Climate Center, Florida State University, Center for Ocean-Atmospheric Prediction Studies), accessed December 8, 2013, http://climatecenter.fsu.edu/products -services/data/1981–2010-normals/apalachicola.

6. The Apalachicola region is so enriched with life that in the middle of the twentieth century Elvy Edison Callaway, a lawyer and Baptist in Florida, considered it the Garden of Eden. Callaway also gave Florida torreya another common name, gopherwood. "Gopher wood" appears in a passage in the Old Testament in reference to the construction of Noah's Ark, though what that wood was supposed to have been is unclear. It has been interpreted variously as boxwood, cedar, or cypress, among certain other woody plants. Callaway, however, used the term "gopherwood" to describe

Florida torreya, and he popularized it as such. He believed that the Ark was built in the Apalachicola region and that during the great flood Noah sailed it across the Atlantic, all the way to Mount Ararat in Armenia. Callaway went so far as to open a Garden of Eden Park on his property along the Apalachicola River in 1956, which included Florida torreya habitat and an entry fee. Callaway's story is recounted in detail in Brook Wilensky-Lanford, *Paradise Lust: Searching for the Garden of Eden* (New York: Grove Press, 2011).

7. The common name "nutmeg" describes the appearance of the seeds, which are nutmeg-like.

8. *Torreya*'s closest relative at the genus level is *Amentotaxus*, a group that contains five conifer species endemic to Asia.

9. Andrew Lee Maxwell, "Mapping and Habitat Analysis of the California Endemic Tree, *Torreya californica*, in Marin County" (master's thesis, San Francisco State University, 1992), offers an overview of the evolutionary history of the *Torreya* genus. An early discussion of *Torreya* distribution is Rudolf Florin, *The Distribution of Conifer and Taxad Genera in Time and Space* (Uppsala: Almqvist and Wiksell, 1963), which indicated that *Torreya* fossils dating to the Lower, or Early, Cretaceous had been unearthed in the New World.

10. The phylogenetics of *Torreya* were explored by Jianhua Li et al., "Phylogenetic Relationships of *Torreya* (Taxaceae) Inferred from Sequences of Nuclear Ribosomal DNA ITS Region," *Harvard Papers in Botany* 6.1 (2001): 275–81.

11. Ibid. The timing of extinction of *Torreya* in Europe is based on the implication that during the Late Tertiary and Quaternary, many plant species in Europe underwent extinction. See also B. H. Tiffney, "Perspectives on the Origin of the Floristic Similarity between Eastern Asia and Eastern North America," *Journal of the Arnold Arboretum* 66 (1985): 73–94; and Qiu-Yun Xiang, Douglas E. Soltis, and Pamela S. Soltis, "The Eastern Asian and Eastern and Western North American Floristic Disjunction: Congruent Phylogenetic Patterns in Seven Diverse Genera," *Molecular Phylogenetics and Evolution* 10.2 (October 1998): 178–90.

12. An early characterization of Florida torreya was provided by A. W. Chapman, "*Torreya taxifolia*, Arnott: A Reminiscence," *Botanical Gazette* 10.4 (April 1885): 251–54. Florida torreya was given various scientific names by botanists, including *Caryotaxus montana*, *C. taxifolia*, *Foetotaxus montana*, and *Tumion taxifolium*. Some of those names are considered to be synonymous with *Torreya taxifolia*, though preference is for the latter.

13. By "fruit," Gray must have been referring to the seeds, which are fairly large and bear a fleshy seed coat.

14. The unusual bijugate trait of *Torreya* species is discussed in detail in Barry P. Tomlinson and Elizabeth H. Zacharias, "Phyllotaxis, Phenology and Architecture in *Cephalotaxus, Torreya and Amentotaxus* (Coniferales)," *Botanical Journal of the Linnean Society* 135.3 (March 2001): 215–28. Interestingly, members of the closely related *Amentotaxus* have a constant decussate arrangement.

15. What information is known on the life history of Florida torreya comes primarily from specimens that have been planted and observed outside the Apalachicola region. The vast majority of those trees have been carefully tended by humans. It is difficult to know whether or how extensively observations of those trees apply to wild Florida torreya, but they do provide at least some approximation. A summary of Florida

torreya's life history is Richard Stalter, "*Torreya taxifolia* Arn. Florida torreya," in *Silvics of North America*, vol. 1, no. 654, ed. Russell M. Burns and Barbara H. Honkala (Washington, DC: US Department of Agriculture, Forest Service, 1990).

16. The possible role of squirrels and tortoises in Florida torreya seed dispersal from Connie Barlow and Paul S. Martin, "Bring *Torreya taxifolia* North—Now," *Wild Earth*, Fall/Winter 2004–5, 55.

17. Documentation of Nieland's observation was cited as personal communication in S. A. Alfieri, Jr., A. P. Martinez, and C. Wehlburg, "Stem and Needle Blight of Florida Torreya, *Torreya taxifolia* Arn.," *Florida State Horticultural Society* 80 (1967): 429.

18. H. Kurz and R. Godfrey, "The Florida Torreya Destined for Extinction," *Science* 136.3,519 (June 1962): 900–902.

19. Alfieri, Jr., Martinez, and Wehlburg, "Stem and Needle Blight."

20. See S. A. Alfieri, Jr., et al., *Index of Plant Diseases in Florida*, Bulletin 11 (Florida Department of Agriculture and Consumer Services, Division of Plant Industry, 1984); and Alfieri, Jr., et al., *Leaf and Stem Disease of* Torreya taxifolia *in Florida*, Plant Pathology Circular no. 291 (Florida Department of Agriculture and Consumer Services, Division of Plant Industry, 1987).

21. The new *Fusarium* species was described in Jason A. Smith et al., "A Novel *Fusarium* Species Causes a Canker Disease of the Critically Endangered Conifer, *Torreya taxifolia*," *Plant Disease* 95.6 (June 2011): 633–39; and Takayuki Aoki et al., "*Fusarium torreyae* sp. nov.: A Novel Pathogen Causing Canker Disease of Florida Torreya (*Torreya taxifolia*), a Critically Endangered Conifer Restricted to Northern Florida and Southwestern Georgia," *Mycologia* 12.262 (October 2012): 312–19.

22. Fungi and oomycetes with possible roles in Florida torreya survival from email communication between the author and Vivian Negrón-Ortiz, 14 January 2014. See also Lydia I. Rivera Vargas and Vivian Negrón-Ortiz, "Root and Soil-Borne Oomycetes (Heterokontophyta) and Fungi Associated with the Endangered Conifer, *Torreya taxifolia* Arn. (Taxaceae) in Georgia and Florida, USA," *Life: The Excitement of Biology* 1.4 (December 2013): 202–23.

23. Somatic embryogenesis for Florida torreya has been explored extensively by plant pathologist Jerry Pullman et al., "Somatic Embryogenesis, Plant Regeneration, and Cryopreservation for *Torreya taxifolia*, a Highly Endangered Coniferous Species," *In Vitro Cellular and Developmental Biology—Plant* 48.3 (June 2012): 324–34.

24. Barlow and Martin, "North—Now," 54.

25. The impact of invasions arising from intracontinental assisted migration from Jillian M. Mueller and Jessica J. Hellmann, "An Assessment of Invasion Risk from Assisted Migration," *Conservation Biology* 22.3 (June 2008): 562–67.

Fickeisen Plains Cactus and Acuña Cactus

1. The common name is pronounced "fik-ai-sin." The taxonomic ranking of Fickeisen plains cactus has been subject to change. At the time of the species' consideration for listing in 1980, it was considered to be a variety (var.). Research since has suggested that it may be a subspecies, *Pediocactus peeblesianus* spp. *fickeiseniae*. Fickeisen plains cactus has been known by other Latin names, including *Navajoa*

fickeisenii, Toumeya fickeisenii, Navajoa peeblesiana ssp. *fickeisenii, Navajoa peeblesiana* var. *fickeisenii,* and *Pediocactus peeblesianus* var. *fickeisenii.*

2. Lyman D. Benson, *The Cacti of Arizona,* 3rd ed. (Tucson: University of Arizona Press, 1969).

3. Data on plant size and flower and fruit production for Fickeisen plains cactus from Lee E. Hughes, "Demographic Monitoring of *P. peeblesianus* var. *fickeiseniae* on the Arizona Strip" (Rocky Mountain Forest and Range Experiment Station, US Department of Agriculture, 1996).

4. For *Pediocactus* species as relicts, see Benson, *Cacti.*

5. Past climatic and floristic features of the Southwest have been surmised from studies of Pleistocene pack rat middens, tree rings, cave sediments, and lake cores. G. Robert Brakenridge estimated regional cooling of 7 to 8°C in the Southwest and surmised a lower timberline, some 2,300 to almost 3,300 feet below current timberline ("Evidence for a Cold, Dry Full-Glacial Climate in the American Southwest," *Quaternary Research* 9.1 [January 1978]: 22–40). A summary of findings is presented in Paul A. Delcourt and Hazel R. Delcourt, "Paleoclimates, Paleovegetation, and Paleofloras of North America North of Mexico during the Late Quaternary," in *Flora of North America North of Mexico,* ed. Flora of North America Editorial Committee (New York: Oxford University Press, 1993), 1:71–94.

6. Information on modern climate change in the Southwest drawn from Gregg Garfin et al., eds., *Assessment of Climate Change in the Southwest United States: A Report Prepared for the National Climate Assessment* (Washington, DC: Island Press, 2013).

7. The link between flower production and water availability in acuña cactus is described in Robert A. Johnson, "Pollination and Reproductive Ecology of Acuña Cactus, *Echinomastus erectrocentrus* var. *Acunensis* (Cactaceae)," *International Journal of Plant Sciences* 153.3, pt. 1 (September 1992): 400–408.

8. Under the Endangered Species Act, the term *critical habitat* refers to specific areas within a species' geographical range, or in some cases outside of its range, that contain physical or biological features that are "essential to the conservation of the species."

Penstemon

1. Louis Krautter, in his Ph.D. thesis, "A Comparative Study of the Genus *Penstemon*" (1908), speculated that Mitchell's description most likely came from observations of material of *P. hirsutus.* It is possible that the plant Mitchell described was already known in botanical terms, possibly having been placed in either of the genera *Chelone, Digitalis,* or *Dracocephalum.*

2. Thomas Nuttall, *The Genera of North American Plants: And a Catalogue of the Species, to the Year 1817* (Philadelphia, 1818, vol. 1; reprint, London: Forgotten Books, 2013), 373–74.

3. Steve L. O'Kane, Jr., and John L. Anderson, "*Penstemon Debilis* (Scrophulariaceae): A New Species from Colorado Endemic to Oil Shale," *Brittonia* 39.4 (October–December 1987): 412–16.

4. Continental radiation in *Penstemon* is described in Andrea D. Wolfe et al., "Phylogeny, Taxonomic Affinities, and Biogeography of *Penstemon* (Plantaginaceae) Based on ITS and cpDNA Sequence Data," *American Journal of Botany* 93.11 (November 2006): 1699–1713.

5. Multiple studies have been carried out that point toward a Rocky Mountain origin of *Penstemon*. From oldest to more recent, the key studies are Richard M. Straw, "A Redefinition of *Penstemon* (Scrophulariaceae)," *Brittonia* 18 (January 1966): 80–95; Andrea D. Wolfe, Shannon L. Datwyler, and Christopher P. Randle, "A Phylogenetic and Biogeographic Analysis of the Cheloneae (Scrophulariaceae) Based on ITS and *matK* Sequence Data," *Systematic Botany* 27 (2002): 138–48; Shannon L. Datwyler and Andrea D. Wolfe, "Phylogenetic Relationships and Morphological Evolution in *Penstemon* subg. *Dasanthera* (Veronicaceae)," *Systematic Botany* 29.1 (2004): 165–76; and Wolfe et al., "Phylogeny, Taxonomic Affinities."

6. *Penstemon* was classified in family Scrophulariaceae, but DNA sequence analyses warranted its transfer to Plantaginaceae. D. C. Albach, H. M. Meudt, and B. Oxelman, "Piecing Together the 'New' Plantaginaceae," *American Journal of Botany* 92.2 (February 2005): 297–315. Hypotheses for the biogeography of *Penstemon* generally are based on the early evolutionary radiation of *Dasanthera*.

7. The Smith Gulch population was later considered as part of the Mount Callahan Saddle population, since the plants inhabited shale at the base of cliffs just below the saddle.

8. Data on remaining Parachute penstemon populations drawn from US Fish and Wildlife Service, "Endangered and Threatened Wildlife and Plants: Determination of Endangered Status for *Ipomopsis polyantha* (Pagosa Skyrocket) and Threatened Status for *Penstemon debilis* (Parachute Beardtongue) and *Phacelia submutica* (DeBeque Phacelia); Final Rule," *Federal Register* 76.14 (27 July 2011).

9. Ibid.

10. Christopher D. Kassotis et al., "Estrogen and Androgen Receptor Activities of Hydraulic Fracturing Chemicals and Surface and Ground Water in a Drilling-Dense Region," *Endocrinology* 155.3 (March 2014): 897–907.

11. Ibid.

12. Pascale Henner et al., "Phytotoxicity of Ancient Gaswork Soils: Effect of Polycyclic Aromatic Hydrocarbons (PAHs) on Plant Germination," *Organic Geochemistry* 30.8 (1999): 963–69.

13. Supposed or demonstrated adverse affects were described in the 2011 FWS Final Rule for Parachute penstemon.

14. Thomas G. Sanders and Jonathan Q. Addo, *Effectiveness and Environmental Impact of Road Dust Suppressants* (Fort Collins, CO: Colorado State University, Department of Civil Engineering, 1993). The US Fish and Wildlife Service Final Rule for *Penstemon debilis* cites an even higher rate of dust deposition: "For every vehicle traveling 1 mi (1.6 km) of unpaved roadway once a day, every day for a year, approximately 2.5 tons of dust are deposited along a 1,000-ft (305–m) corridor centered on the road" (*Federal Register* 76.144 [27 July 2011], 45064).

15. By 2011 another third of Parachute penstemon habitat was managed by the BLM to minimize certain threats. The remaining habitat and populations, all on oil company lands, were unprotected.

16. For example, while no surface occupancy prevented the development of drilling pads, pipelines could be constructed, since disturbed areas could be revegetated.

17. The Rocky Mountain Cliff and Canyon plant community is characterized by sparse vegetation, which may include scattered shrubs and limited herbaceous and tree cover.

18. For the final rule on critical habitat, see US Fish and Wildlife Service, "Designation of Critical Habitat for *Ipomopsis polyantha* (Pagosa skyrocket), *Penstemon debilis* (Parachute beardtongue), and *Phacelia submutica* (DeBeque phacelia); Final Rule," *Federal Register* 77.156 (13 August 2012). Additional critical habitat features for Parachute penstemon are discussed in US Fish and Wildlife Service, "Recovery Outline: Parachute Beardtongue (*Penstemon debilis*)" (Western Colorado Ecological Services Field Office, January 2013). Because of Oxy's prominent role in protecting the species, Oxy lands were excluded from the 2012 final rule on critical habitat.

19. "Final Supplemental Environmental Impact Statement for the Prototype Oil Shale Leasing Program" (US Bureau of Land Management, Colorado State Office, 1983), accessed 20 December 2013, https://archive.org/details/finalsupplementa03unit.

Golden Paintbrush

1. Golden paintbrush can be confused with harsh Indian paintbrush (*Castilleja hispida*), which occasionally produces yellow bracts. Most of the time, harsh Indian paintbrush produces scarlet bracts. The species is also found in areas far beyond the range of golden paintbrush.

2. S1 is one of multiple conservation status rankings devised by NatureServe. S indicates the geographic scale of "state." It is one of three distinct geographic scales used, with N (national) and G (global) being the other two. The number 1 designates a species as critically imperiled, whereas a ranking of 5 designates it as secure. The letters X and H are used in place of number rankings in cases where a species is presumed to be extinct or extirpated (X) or is known only from historical records (H) and is therefore likely to be extinct or extirpated.

3. The population at San Juan Valley consisted of 7,528 plants in 2003, but the site was on private property, and the owners denied access for later surveys. As a result, whether that number still held several years later was unknown.

4. Information on collectors and years and sites of collection drawn primarily from the Consortium of Pacific Northwest Herbaria, http://www.pnwherbaria.org.

5. J. M. Greenman, "Some New and Other Noteworthy Plants of the Northwest," *Botanical Gazette* 25 (1 April 1898): 261–69. Greenman later was known for his tenure (1913–48) as botanist and curator of the herbarium at the Missouri Botanical Garden.

6. Loss of potential golden paintbrush habitat from construction of the Bonneville Lock and Dam is mentioned in Beth A. Lawrence and Thomas N. Kaye, "Habitat Variation throughout the Historic Range of Golden Paintbrush, a Pacific Northwest Prairie Endemic: Implications for Reintroduction," *Northwest Science* 80.2 (2006): 140–52. Lawrence and Kaye suspect that "Peterson Butte Cemetery" is in fact Sand Ridge Cemetery, which sits at the base of Peterson's Butte.

7. T. N. Kaye and B. A. Lawrence, "Fitness Effects of Inbreeding and Outbreeding on Golden Paintbrush (*Castilleja levisecta*): Implications for Recovery and Reintroduction" (Institute for Applied Ecology, Corvallis, Oregon, and Washington Department of Natural Resources, Olympia, Washington, 2003).

8. COSEWIC, *COSEWIC Assessment and Update Status Report on the Golden Paintbrush* Castilleja levisecta *in Canada* (Committee on the Status of Endangered Wildlife in Canada, Ottawa, 2007).

9. B. A. Lawrence and T. N. Kaye, "Growing *Castilleja* for Restoration and the Garden," *Rock Garden Quarterly* 63 (2005): 128–34.

10. Lynn S. Adler, "Host Species Affects Herbivory, Pollination, and Reproduction in Experiments with Parasitic *Castilleja*," *Ecology* 84.8 (August 2003): 2083–91.

11. There is some debate as to whether red fescue is a native species or an introduced species in the Puget Trough province. Data on vegetation cover in golden paintbrush habitat from Lawrence and Kaye, "Habitat Variation," 145.

12. Trial Island and Alpha Islet are now fully protected within ecological reserves—respectively, the Trial Islands Ecological Reserve and the Oak Bay Islands Ecological Reserve.

13. Lawrence and Kaye, "Habitat Variation," 145–46.

14. Ted Thomas and John Gamon, "Restoration of a Prairie Plant Community: Help for a Threatened Species," in *Proceedings of the Fifteenth North American Prairie Conference*, ed. Charles Warwick (1997), 244–48.

15. The first US recovery strategy was John Gamon, Peter Dunwiddie, and Ted Thomas, "Recovery Plan for the Golden Paintbrush (*Castilleja levisecta*)" (US Fish and Wildlife Service, Portland, Oregon, 2000). The recovery strategy for maritime meadows in Canada is described in Parks Canada Agency, "Recovery Strategy for Multispecies at Risk in Maritime Meadows Associated with Garry Oak Ecosystems in Canada," in *Species at Risk Act*, Recovery Strategy Series (Ottawa: Parks Canada Agency, 2006), http://www.sararegistry.gc.ca/virtual_sara/files/plans/rs_maritime _meadow_0806_E.pdf.

16. Parks Canada Agency, "Recovery Strategy," v, 2.

17. Dennis Aubrey, "Oviposition Preference in Taylor's Checkerspot Butterflies (*Euphydryas editha taylori*): Collaborative Research and Conservation with Incarcerated Women" (master's thesis, Evergreen State College, June 2013).

18. See Gamon, Dunwiddie, and Thomas, "Recovery Plan."

19. US Fish and Wildlife Service, "Golden Paintbrush 5-Year Review" (US Fish and Wildlife Service Western Washington Fish and Wildlife Office, Lacey, Washington, 2007).

20. Mary Jo W. Godt, Florence Caplow, and J. L. Hamrick, "Allozyme Diversity in the Federally Threatened Golden Paintbrush, *Castilleja levisecta* (Scrophulariaceae)," *Conservation Genetics* 6 (2005): 87–99.

21. Peter W. Dunwiddie, R. Adam Martin, and Marion Cady Jarisch, "Water and Fertilizer Effects on the Germination and Survival of Direct-Seeded Golden Paintbrush (*Castilleja levisecta*)," *Ecological Restoration* 31.1 (March 2013): 10–12. Possible delisting by 2018 via email communication between the author and Tom Kaye.

22. B. A. Lawrence and T. N. Kaye, "Reintroduction of *Castilleja levisecta*: Effects of Ecological Similarity, Source Population Genetics, and Habitat Quality," *Restoration Ecology* 19.2 (March 2011): 166–76.

Hidden Value

1. Batelle Technology Partnership Practice, *Economic Impact of the Human Genome Project* (Batelle Memorial Institute, 2011). Some economists have argued that $796 billion is an overestimate.

2. Two key whole-genome sequencing studies of sorghum are A. H. Patterson et al., "The *Sorghum bicolor* Genome and the Diversification of Grasses," *Nature* 457.7,229 (29 January 2009): 551–56; and Emma S. Mace et al., "Whole-Genome Sequencing Reveals Untapped Genetic Potential in Africa's Indigenous Cereal Crop Sorghum," *Nature Communications* 4 (August 2013), doi:10.1038/ncomms3320.

3. David E. Cook et al., "Copy Number Variation of Multiple Genes at *Rhg1* Mediates Nematode Resistance in Soybean," *Science* 338.6,111 (November 2012): 1206–9.

4. The early demonstration of transformation of rice was described in Ko Shimamoto et al., "Fertile Transgenic Rice Plants Regenerated from Transformed Protoplasts," *Nature* 338 (16 March 1989): 274–76. The *hph* gene made plants resistant to hygromycin B, a cell-killing antibiotic produced by *Streptomyces hygroscopicus* bacteria.

5. The genetic diversity of Texas wild rice stands is described in Christopher M. Richards et al., "Capturing Genetic Diversity of Wild Populations for *Ex Situ* Conservation: Texas Wild Rice (*Zizania texana*) as a Model," *Genetic Resources and Crop Evolution* 54 (2007): 837–48.

6. M. Abedinia et al., "Accessing Genes in the Tertiary Gene Pool of Rice by Direct Introduction of Total DNA from *Zizania palustris* (Wild Rice)," *Plant Molecular Biology Reporter* 18 (2000): 133–38.

7. As projected by David Tilman et al., "Global Food Demand and the Sustainable Intensification of Agriculture," *Proceedings of the National Academy of Sciences* 108.50 (13 December 2011): 20260–64.

8. *Indica* and *japonica* cultivars are grown in the United States.

9. Daisuke Fujita et al., "*NAL1* Allele from a Rice Landrace Greatly Increases Yield in Modern *Indica* Cultivars," *Proceedings of the National Academy of Sciences* 110.51 (17 December 2013): 20431–36.

10. Shaobing Peng et al., "Rice Yields Decline with Higher Night Temperature from Global Warming," *Proceedings of the National Academy of Sciences* 101.27 (6 July 2004): 9971–75.

11. A detailed account of the relationship of the Ojibwe with wild rice is Thomas Vennum, *Wild Rice and the Ojibway People* (St. Paul: Minnesota Historical Society Press, 1988).

12. Whether herbal products that are based on *Echinacea* extracts are actually effective remains a matter of debate. Most scientific investigations that have focused on extracts have been inconclusive or have presented conflicting results—variations attributed to inconsistencies in the preparation and standardization of the products studied.

13. Inhibition of LPS-induced nitric oxide production from Ying Chen et al., "Macrophage Activating Effects of New Alkamides from the Roots of *Echinacea* Species," *Journal of Natural Products* 68.5 (2005): 773–76. Inhibition of inflammatory mediators from B. Müller-Jakic et al., "In Vitro Inhibition of Cyclooxygenase

and 5-Lipoxygenase by Alkamides from *Echinacea* and *Achillea* Species," *Planta Medica* 60.1 (February 1994): 37–40.

14. Annie Cheminat et al., "Caffeoyl Conjugates from *Echinacea* Species: Structures and Biological Activity," *Phytochemistry* 27.9 (1988): 2787–94.

15. Chicoric acid takes its name from chicory, the species from which it was first isolated. M. L. Scarpati and G. Oriente, "Chicoric Acid (Dicaffeyltartic Acid): Its Isolation from Chicory (*Chicorium intybus*) and Synthesis," *Tetrahedron* 4.1–2 (1958): 43–48.

16. For interference with HIV, see Wim Pluymers et al., "Viral Entry as the Primary Target for the Anti-HIV Activity of Chicoric Acid and Its Tetra-Acetyl Esters," *Molecular Pharmacology* 58.3 (1 September 2000): 641–48.

17. Jae Yeol Lee, Kwon Joong Yoon, and Yong Sup Lee, "Catechol-Substituted L-Chicoric Acid Analogues as HIV Integrase Inhibitors," *Bioorganic & Medicinal Chemistry Letters* 13 (2003): 4331–34.

18. Per Molgaard et al., "HPLC Method Validated for the Simultaneous Analysis of Cichoric Acid and Alkamides in *Echinacea purpurea* Plants and Products," *Journal of Agricultural and Food Chemistry* 51 (2003): 6922–33.

19. Pacific Yew Act of 1992, H.R. 3836, 102nd Cong. (1992).

The Northern Plants

1. In some regions, particularly in Canada, the term *boreal* may be used to refer to the southerly, vegetated parts of the biome, while *taiga* is used to describe northerly, barren areas.

2. The definition of "northern" used here is for convenience; it is not a technical delineation.

3. Multiple studies have explored the vulnerability of various of the world's biomes to climate change. See, in particular, Patrick Gonzalez et al., "Global Patterns in the Vulnerability of Ecosystems to Vegetation Shifts Due to Climate Change," *Global Ecology and Biogeography* 19 (2010): 755–68; and Marko Scholze et al., "A Climate-Change Risk Analysis for World Ecosystems," *Proceedings of the National Academy of Sciences* 103.35 (29 August 2006): 13116–20. The predicted impact of climatic warming on subpolar regions is described in William L. Chapman and John E. Walsh, "Recent Variations of Sea Ice and Air Temperature in High Latitudes," *Bulletin of the American Meteorological Society* 74.1 (January 1993): 33–47; and IPCC, *Climate Change 2001, the Scientific Basis: Summary for Policymakers. A Report of Working Group 1 of the Intergovernmental Panel on Climate Change* (New York: Cambridge University Press, 2001).

4. Several studies have pointed toward such an outcome. See Robert T. Watson, Marufu C. Zinyowera, and Richard H. Moss, eds., *IPCC Special Report on the Regional Impacts of Climate Change: An Assessment of Vulnerability* (New York: Cambridge University Press, 1998); and Charles D. Koven, "Boreal Carbon Loss Due to Poleward Shift in Low-Carbon Ecosystems," *Nature Geoscience* 6 (2013): 452–56.

5. Estimates on boreal carbon storage have varied. The range provided here is from Yude Pan et al., "A Large and Persistent Carbon Sink in the World's Forests,"

Science 333 (19 August 2011): 988–93; and Eric S. Kasischke, "Boreal Ecosystems in the Global Carbon Cycle," in *Fire, Climate Change, and Carbon Cycling in the Boreal Forest*, ed. Eric S. Kasischke et al. (New York: Springer-Verlag, 2000).

6. Climate change itself is not uniform across northern regions. For example, some areas are experiencing more significant changes in precipitation than others. An example of synergism between ecological threats is the relationship between invasive species and climatic warming. An increasingly early arrival of spring, for instance, favors the emergence of invasive plants, which tend to be more adaptable than native species to changing climatic conditions.

7. Along with all the other members of *Pedicularis*, Furbish lousewort was once classified in the snapdragon family (Scrophulariaceae). The genus, which contains several hundred species, is now placed in the broomrape family (Orobanchaceae).

8. Information on the pollination and sexual reproduction of Furbish lousewort drawn from US Fish and Wildlife Service, "Revised Furbish Lousewort Recovery Plan" (Newton Corner, Massachusetts, 1991). Lazarus Walter Macior, "The Pollination Ecology and Endemic Adaptation of *Pedicularis furbishiae* S. Wats," *Bulletin of the Torrey Botanical Club* 105.4 (October–December 1978): 268–77, was also consulted, though more recent studies have rendered some of Macior's findings obsolete. In his 1978 paper, Macior noted that other species of *Pedicularis* that inhabit temperate regions relied on either queens or queens and workers for pollination.

9. D. M. Waller, D. M. O'Malley, and S. C. Gawler, "Genetic Variation in the Endemic *Pedicularis furbishiae* (Scrophulariaceae)," *Conservation Biology* 1.4 (December 1987): 335–40.

10. The first to collect Furbish lousewort may have been Canadian naturalist John Moser in 1878, at Grand Falls in New Brunswick. Hay likely collected it the following year. However, both men apparently mistook the plant for common lousewort (*P. canadensis*), leaving Kate Furbish's find in 1880 to be the one that led to the discovery of the plant's true identity. "Hoping to add a few new species," as recounted in Kate Furbish, "A Botanist's Trip to 'The Aroostook,'" *American Naturalist* 15 (June 1881): 469–70, 469. Also quoted in Macior, "The Furbish Lousewort: Weed, Weapon, or Wonder?," *American Biology Teacher* 43.6 (September 1981): 323–26.

11. Furbish, "A Botanist's Trip," 470.

12. Secretary, Smithsonian Institution, *Report on Endangered and Threatened Species of the United States*, Committee on Merchant Marine and Fisheries, Serial No. 94-A, 94th Cong., 1st sess., House Document 94–51 (Washington, DC: US Government Printing Office, 1975).

13. Data from "90-Day Finding on a Petition to Delist *Pedicularis furbishiae* (Furbish lousewort) and Initiation of a 5-Year Status Review," *Federal Register* 70.153 (10 August 2005).

14. Data on 1980 population counts in Maine and New Brunswick from SARA Public Registry, "Species Profile: Furbish's Lousewort," 2008. Data on 1989 count in Maine from US Fish and Wildlife Service, "90-Day Finding." Size of Canadian population in 2011 from COSEWIC, *COSEWIC Status Appraisal Summary on Furbish's Lousewort* Pedicularis furbishiae *in Canada* (Committee on the Status of Endangered Wildlife in Canada, Ottawa, 2011).

15. Furbish's Lousewort Recovery Team, "Recovery Strategy for Furbish's lousewort (*Pedicularis furbishiae*) in New Brunswick" (New Brunswick Department of Natural Resources, Fredericton, New Brunswick, 2006).

16. Slight warming trend along Saint John River from Spyros Beltaos, "Climatic Effects on the Changing Ice-Breakup Regime of the Saint John River," in *River Ice Management with a Changing Climate: Dealing with Extreme Events, Proceedings of the Tenth Workshop on River Ice, Winnipeg, Canada* (Committee on River Ice Processes and the Environment, Hydrology Section, Canadian Geophysical Union, 1999). Research on climate change and the Saint John River is summarized in Mark McCollough, US Fish and Wildlife Service, "Furbish's lousewort (*Pedicularis furbishiae*) Five-Year Review: Summary and Evaluation" (US Fish and Wildlife Service, Maine Field Office, Old Town, Maine, 2007).

17. As recorded in Sir William Jackson Hooker, *Flora Boreali–Americana: Or, the Botany of the Northern Parts of British America,* vol. 1 (London: H. G. Bohn, 1840). Hooker describes the location of Richardson's find as at the mouth of the Mackenzie River. It is more likely, however, that the specimen Hooker described came from Cape Bathurst. A possible explanation for the type locality given by Hooker is presented in James G. Harris, "Pilose Braya, *Braya pilosa* Hooker (Cruciferae; Brassicaceae), an Enigmatic Endemic of Arctic Canada," *Canadian Field-Naturalist* 118.4 (2004): 550–57.

18. Walter N. Meier, Julienne Stroeve, and Florence Fetterer, "Whither Arctic Sea Ice? A Clear Signal of Decline Regionally, Seasonally and Extending beyond the Satellite Record," *Annals of Glaciology* 46.1 (2007): 428–34; and Donald K. Perovich and Jacqueline A. Richter-Menge, "Loss of Sea Ice in the Arctic," *Annual Review of Marine Science* 1 (2009): 417–41.

19. Detailed information on hairy braya subpopulations and their distribution is provided in "Species Status Report for Hairy Braya (*Braya pilosa*) in the Northwest Territories" (Canadian Species at Risk Committee, 2012). In the report, a subpopulation is defined as a group of plants separated from another group by at least 1 kilometer, with the area between being made up of unsuitable habitat. Much of what is known about those subpopulations is derived from research carried out by the report's author, biologist James G. Harris, who is the foremost expert on hairy braya. Because nearly all of what has been published about hairy braya in recent years comes from Harris's research, much of what is presented here relates the story of his remarkable effort to better understand the mysterious Arctic plant.

20. The close relationship between Greenland northern rockcress and hairy braya, based on unpublished genetic findings, per email communication between the author and James G. Harris, 1 July 2014.

21. Coastal erosion rate for the region drawn from the 2012 Canadian Species at Risk Committee report and from Benjamin M. Jones et al., "Increase in the Rate and Uniformity of Coastline Erosion in Arctic Alaska," *Geophysical Research Letters* 36.L03503 (February 2009).

22. For persistence of climate change, see Susan Solomon et al., "Irreversible Climate Change Due to Carbon Dioxide Emissions," *Proceedings of the National Academy of Sciences* 106.6 (10 February 2009): 1704–09; and IPCC, "Summary for Policymakers," in *Climate Change 2013: The Physical Science Basis. Contribution of Working Group I to the Fifth Assessment Report of the Intergovernmental Panel on Climate Change,* ed. T. F. Stocker et al. (New York: Cambridge University Press, 2013).

Afterword

1. Data on percentages of North American plant species in ex situ collections from Abbey Hird and Andrea T. Kramer, "Achieving Target 8 of the Global Strategy for Plant Conservation: Lessons Learned from the North American Collections Assessment," *Annals of the Missouri Botanical Garden* 99.2 (2013): 161–66; A. Kramer et al., *Conserving North America's Threatened Plants: Progress Report on Target 8 of the Global Strategy for Plant Conservation* (Botanic Gardens Conservation International US, Glencoe, Illinois, 2011); and BGCI US, United States Botanic Garden, *Progress Report on Target 8 of the Global Strategy for Plant Conservation in the United States* (Botanic Gardens Conservation International US, Glencoe, Illinois, 2014).

2. In addition to the GSPC framework, the North American Collections Assessment also took into account goals set out by other plant conservation efforts, including the International Agenda for Botanic Gardens in Conservation and the National Framework for Progress in Plant Conservation developed by the Plant Conservation Alliance.

3. Target 8 is just one of sixteen GSPC targets.

Index

Page numbers in **bold** indicate illustrations.

torchwood family, 55
Torrey, John, 13, 80, 85, 96–97, 218n2
Torreya: Amentotaxus and, 220n8,
220n14; divergence of, 95; geographic
distribution of, 94–95, 106, 220n9;
migration of, 95–96, 107; species of,
94. *See also* Florida torreya
Torreya Guardians, 106, 107, 109
Torreya State Park, 101, 102, 104, 105, 110
Torreya taxifolia. See Florida torreya
Travels (Bartram), 60, 62, 65, 217n4
tree islands, 11
tree mortality, 25, 214n11
Trial Island, 149, 157, 158, 159, 160, 166,
167, 169, 225n12
Trifolium reflexum, 216n8
Trifolium stoloniferum. See running
buffalo clover
Tuktoyaktuk Community Conservation
Plan, 205
tundra, 37, 118, 187–88, 202, 205, 215n6
tundra plains, 189, 202, 205
Turgai Sea, 95

Upton Garden, 60
US Army Corps of Engineers, 73, 102,
104, 197
US Bureau of Land Management
(BLM): Fickeisen plains cactus and,
112, 113, 116, 121, 122, 123, 124–26;
Pacific yew and, 186; Parachute
penstemon and, 138, 144, 146, 223n15
US Fish and Wildlife Service, 10, 50, 73,
81–82, 86, 87, 104, 168
US Forest Service, 24, 45, 142, 186

Vancouver Island: development of, 151,
152; Garry oak ecosystem on, 152, 165,
169; geography of, 149; golden
paintbrush and, 148, 152, 153, 157,
170; settlement of, 151
Voyageurs National Park, 41
Vroom, James, 195–96

Waterton Lakes National Park, **12**, 18
Watson, Sereno, 196
westward expansion, 76, 89
Wetmore, J. E., 195
Wherry, Edgar T., 64
Whidbey Island, 149, 158, 161, 163, 166,
169, 170
whitebark pine: climate change and, 9,
21, 22, 24; component of mixed stands,
13–14; fire and regeneration of, 21–22;
genetic history, diversity of, 17, 23, 24;
loss of, 9, 10, 22–23; range of, 13; rate
of growth, cones of, 14–15, **16**, 18;
restoration of, 17, 23–24; role in
subalpine, 9, 10–11, 13, 18, 22; seeds,
seed dispersal of, 11, 15, 17–18, 22,
213n3; at tree line, 11, 13, 14, 23.
See also Clark's nutcracker; mountain
pine beetle; white pine blister rust
white pine, 9, 11, 17, 19, 21, 22,
41, 43
white pine blister rust: 9, 17, 18–19, 21,
22, 23–24
White River beardtongue, 146
wildfire, 6, 25, 44, 121
wild medicinal plants, 182
wild rice, 175–77, 178–81, 187, 226n5
Wilhelm, Gerould, 73
Willamette Valley: agriculture in,
151–52, 160; climate of, 158; habitat
loss in, 152, 160; habitat types in,
158, 159; Mill Plain, 153; nonnative
species in, 161, 167; physiographic
province, 157–58; settlement in, 151;
soils of, 157. *See also* golden
paintbrush

Yellowstone, 10, 18, 89
yew family, 97
Young, William, 61
Yuchi, 46

Zizania. See wild rice

About the Author

KARA ROGERS is the senior editor of biomedical sciences at Encyclopædia Britannica, Inc. She holds a B.S. in biology and a Ph.D. in pharmacology and toxicology and is a member of the National Association of Science Writers. She is the author of *Out of Nature: Why Drugs from Plants Matter to the Future of Humanity* (University of Arizona Press, 2012).